Leisure, Plantations, and the Making of a New South

New Studies in Southern History

Series Editor: John David Smith,
The University of North Carolina at Charlotte

Leisure, Plantations, and the Making of a New South

The Sporting Plantations of the South Carolina Lowcountry and Red Hills Region, 1900–1940

Edited by
Julia Brock and Daniel Vivian

LEXINGTON BOOKS
Lanham • Boulder • New York • London

Published by Lexington Books
An imprint of The Rowman & Littlefield Publishing Group, Inc.
4501 Forbes Boulevard, Suite 200, Lanham, Maryland 20706
www.rowman.com

Unit A, Whitacre Mews, 26-34 Stannary Street, London SE11 4AB

British Library Cataloguing in Publication Information Available

Library of Congress Control Number: 2015949365
ISBN: 978-0-7391-9578-9 (cloth : alk. paper)
eISBN: 978-0-7391-9579-6

♾™ The paper used in this publication meets the minimum requirements of American National Standard for Information Sciences—Permanence of Paper for Printed Library Materials, ANSI/NISO Z39.48-1992.

Printed in the United States of America

Contents

Acknowledgments

This collection would not have been possible without the support and guidance of many friends and colleagues. We especially wish to thank Robert Weyeneth, who recognized the connections in our research and introduced us; John David Smith, who reviewed our initial proposal and recommended it to Lexington Books; Brian Hill, Erin Wapole, and Brighid Stone at Lexington Books for their conscientious advice and assistance throughout the editorial process; Albert Way, who supplied advice and encouragement on numerous occasions; and the contributors, all of whom worked earnestly and energetically on their essays and responded to suggestions for revision thoughtfully and enthusiastically. Staff at several organizations also facilitated production of the manuscript. At the South Carolina Historical Society, Mary Jo Fairchild and Virginia Ellison provided prompt and cheerful assistance with image requests. George Chastain of the Belle W. Baruch Foundation and Julie Warren of the Georgetown County Digital Library rendered similar assistance with Jennifer Betsworth's essay. At the Thomas County Historical Society, Ann Harrison and Ephraim Rotter expressed support for the project and generously supplied images for Julia Brock's essay. Jennifer McCormick of the Charleston Museum deserves special thanks for assisting with an image request on short notice for Daniel Vivian's essay.

Colleagues at each of our respective institutions and friends and family also gave valuable support. Tracy K'Meyer showed interest in this project while still in its conceptual stages and provided valuable advice about organization and editing. Jennifer Dickey and Catherine Lewis offered consultation and encouragement throughout the project's development. Hannah O'Daniel, a graduate student at the University of Louisville, showed exceptional skill and attention to detail in proofreading and formatting all of the essays.

Julia Brock and Daniel Vivian

Introduction

Julia Brock and Daniel Vivian

In February 1935, *Country Life* magazine reported on an "army" that had recently "marched through Georgia." Unlike the one that had terrorized white Georgians and sparked jubilation among their slaves in the fall of 1864, this one had "entirely peaceful" aims. With "sport and sanctuary from the rigors of the Northern winter" as primary goals, its members had created "plantations" dedicated to their favored pastimes. "The shooting was good and the climate was ideal," the magazine reported. "Flowers grew in profusion," and warm hospitality—"a symbol of the South"—won the visitors over. Now, *Country Life* observed, "some 30,000 acres of hunting preserves" lay in the territory between Thomasville and Tallahassee, Florida. The "Red Hills"—so named for its undulating terrain and the color of its clay soils—had become a favored destination for wealthy sporting enthusiasts from afar. Each fall and winter, sportsmen and sportswomen from northern cities flocked to the region to enjoy mild weather and good sport.[1]

Some 250 miles to the northeast, similar developments had occurred. All along the coast of South Carolina, wealthy sporting enthusiasts from northern cities passed the winter months at grand estates. By the mid-1930s, more than seventy lay in the territory between Georgetown and the Savannah River, a distance of roughly 130 miles. *Country Life* also highlighted their development. In a January 1932 article, it praised sportsmen and sportswomen for creating estates characterized by "simple dignity" and beauty. As in the Red Hills, most had roots in old plantations. Sportsmen and sportswomen purchased plantations, rehabilitated existing buildings, added new structures,

1

and styled the resulting whole as they saw fit. They came annually to enjoy warm breezes, stunning scenery, and their favorite pastimes.[2]

Country Life's reporting calls attention to a phenomenon that has received little attention from historians. In the decades after Reconstruction, sportsmen from northern cities and enterprising southerners established hunting plantations in select locations across the South. Old plantations adapted for use as private hunting retreats provided access to wildlife, undeveloped land, and lodgings. In many cases, sportsmen showed little concern for refinement. The masculine ethos of hunting militated against aesthetic pretention and material comfort. Yet in a few instances, sportsmen broke from the norm. The "plantations" of the Red Hills region and the Carolina coast—"the lowcountry," in popular parlance—offer two examples. They represent the two most significant instances where wealthy sporting enthusiasts developed large, elegantly styled "plantations." By the 1930s, both regions won widespread acclaim for their role as seasonal destinations for prominent businessmen, bankers, politicians, and heirs of industrial fortunes.[3]

Photographs that appeared in the pages of *Country Life* left no doubt about the character of the new estates. Owners passed the winter months in handsome mansions surrounded by neatly kept lawns and beautiful gardens. Spacious accommodations, majestic surroundings, and large acreages proved typical. At a time when the word "plantation" typically signified a cluster of sharecropping tracts or tenant farms, wealthy sportsmen and sportswomen created examples of a dramatically different kind. With their rise, plantations moved onto new terrain.[4]

The sporting plantations of the Red Hills and lowcountry have long been known to historians and the public alike. In the late 1960s, scholars such as George Brown Tindall, William R. Brueckheimer, Clifton Paisley, William Warren Rogers, and George C. Rogers, Jr. charted their origins and development. They saw the combination of inexpensive land and plentiful wildlife as sufficient to explain the new estates. George Rogers also added the influence of the "plantation myth," which he judged important to newly wealthy families "searching for status." Yet otherwise, Tindall, Bruckheimer, and Rogers viewed the new estates as unremarkable. Although impressed by their scale and sumptuousness, they found them easily explained. Wealthy sporting enthusiasts discovered unspoiled territory with large game populations; from there the rest fell into place.[5]

In the decades since, other scholars have shown occasional interest in the estates of both regions, and a handful of studies have probed select dimensions of their use and influence. Recent investigations have examined the architecture and material development of select examples and the environmental consequences of sportsmen's and sportswomen's activities. These inquiries have illuminated complex processes behind the rise of the new estates and important variations among them. They have called attention to the varied uses that sporting plantations served, the complex roles they assumed in the communities where they developed, and the experiences they provided for owners, guests, and workers.[6] Yet despite these investigations, much about the new estates remain obscure. The sources of their inspiration, the conditions that facilitated their development, and the reasons that sportsmen and sportswomen favored the Red Hills and the lowcountry over other parts of the South remain poorly understood.

Sporting plantations represented a greater innovation than generally recognized. Although some examples amounted to little more than old plantations adapted for new purposes, those of the Red Hills and the lowcountry entailed considerably more. Elegant architecture, well-developed outbuildings, and elaborate landscaping identified them as sites of performance and display. Owners entertained friends, family, and business associates in settings designed to impress. Large acreages identified the new estates as hunting domains on a grand scale. The social activities that owners and guests engaged in also set the plantations of the Red Hills and the lowcountry apart. When *Country Life* referred to "life on the Thomasville-Tallahassee 'circuit,'" it alluded to rituals of house parties, visiting, lunches and dinners, and an endless stream of hunts. Throughout the fall and winter months, estate owners and their guests busied themselves with constant activity. In some cases, owners even complained about relentless socializing. Owning a "southern plantation" promised peace, quiet, and solitude, but at least some members of the new planter class discovered that reality clashed with their expectations.[7]

Discourses associated with the new estates also hint at their significance. Estate owners, their guests, and contemporaries all spoke of them as "plantations" in a manner that ascribed new meanings to the term. All used "estate" and "plantation" interchangeably, virtually without distinction. "Plantation," however, represented more than a holdover from an earlier period. Laden with meaning, it informed views of the estates and their relationship to plantations of earlier periods. It simultaneously indicated ties to the past and evolutionary development. In short, people of the era saw these estates as a new type of plantation, different yet not cut from whole cloth.[8]

How should the sporting plantations of the Red Hills and the lowcountry be understood? Why did they develop, and what did they matter? And how did people of the era understand them? Plantations, after all, had never before served solely as sites of upper-class leisure and recreation. Never before had they served as seasonal residences for members of a social and economic elite. For roughly 500 years, the term "plantation" had denoted a place of agricultural production worked at least in part by slave labor. The plantations worked by sharecroppers and tenants that stretched across the southern

United States represented the post-emancipation variant of a brutally exploi-
tive system of economic and social organization. By the 1930s, observers
near and far saw them as fundamental to the problems that plagued the South,
namely soul-crushing poverty, poor public health, and low levels of education. Sporting plantations represented something markedly different. They
set small numbers of plantations on a different course and, in the process,
opened up new opportunities for small numbers of people, some native to the
region, some not.[9]

This collection investigates the sporting plantations of the Red Hills and
the lowcountry. It seeks to understand the new estates as people who saw and
experienced those in the era of their creation did. It explores their origins and
development, their role in reshaping established conceptions of plantations,
and how they affected the landscape and ecology of both regions. Examining
these topics brings much needed scrutiny to histories that have yet to receive
due attention. Despite growing interest in sporting plantations, scholars have
failed to recognize the complex motivations that lay behind them and the
consequences of their development. Considering the disparate and varied
influences that informed sportsmen's and sportswomen's actions shows the
new estates to have been less the product of good fortune—the "discovery" of
cheap land and plentiful wildlife—than deliberate efforts to secure access to
valued resources. Examining the relationships that estate owners developed
with white and black southerners reveals that the former had less agency than
generally assumed and that relationships of mutual benefit and dependence
predominated. Investigating the influence of the new estates on the lands
they occupied and wildlife populations sets sportsmen's and sportswomen's
conservation efforts in a new perspective. In short, critical analysis of basic
processes associated with the development of the new estates shows them
to have been anything but inevitable. Rather, they resulted from deliberate
choices, struggles over scarce resources, and efforts to attain experiences not
possible elsewhere.

The place to begin inquiry into these topics is with the growth of
recreational hunting in the decades after the Civil War. Although wealthy
Americans hunted for pleasure during the colonial and antebellum eras, the
rapid expansion of industrial wealth afterward gave sport hunting new promi-
nence. As newly wealthy families vied for status, hunting became central to
a peripatetic lifestyle that prized outdoor recreation and social intercourse in
exclusive settings. The sporting plantations of the Red Hills and the lowcountry developed amid these circumstances. Neither country estates nor plantations in the traditional sense of the word, they incorporated elements of both
idioms in pursuit of ideals then in vogue. For small numbers of upper-class
Americans, hunting estates provided entrée to an exclusive realm of activity

that paid homage to aristocratic traditions while embracing the contemporary zeal for outdoor recreation and the status-conscious behaviors of the era.

Sporting plantations developed amid the social and cultural upheavals of the Gilded Age. In the decades following the Civil War, the popularity of sport hunting surged. Precise numbers are elusive, for no single organization tracked hunters through licenses or memberships. Still, extensive evidence demonstrates growing enthusiasm for the pastime. Sporting periodicals became ubiquitous. Between 1865 and 1900, thirty-nine began publication; the most successful attained a circulation of nearly 100,000. Sporting manuals also flooded the market, eager to inform prospective sportsmen about proper sporting practices, the characteristics of particular types of game, and valued traditions. Vacations to remote destinations also became popular. The Adirondack Mountains of New York quickly became a prized destination for middle-class urbanites seeking to restore themselves amid unspoiled nature. Popularized in part by William H. H. "Adirondack" Murray, a Protestant minister whose guidebook, *Adventures in the Wilderness; or, Camp-life in the Adirondacks*, became an overnight sensation when published in 1869, men from New York, Boston, and other northeastern cities turned wilderness outings into journeys of moral and physical renewal. For thousands of Americans, hunting became a means of claiming manhood and warding off the enervating effects of modern life.[10]

Sport hunting's newfound popularity stemmed directly from the growth of an urban middle class. Before the Civil War, only wealthy Americans, mainly merchants and southern planters, hunted for recreation. After Appomattox, middle-class urbanites became the sport's main constituency. The rapid expansion of the American economy created a class of clerks, salesmen, managers, government employees, and salaried professionals with the resources needed to make leisure a regular part of life. At the same time, the conditions of urban life raised concerns about overwork, nervous disorders, the effeminizing effects of "overcivilization," and the loss of direct contact with nature. Hunting supplied what many viewed as an elixir to these problems. Its emphasis on skill and self-control offered opportunities for harried businessmen and professionals to resist temptation and develop discipline. The wholesome, restorative influence of nature offered a powerful counterpoint to the devitalizing effects of urban life. Moreover, the camaraderie that hunting fostered seemed crucial for developing character, the elusive quality that Victorian culture viewed as necessary for success. As urbanization and industrialization transformed the nation, thousands took up hunting with enthusiasm. Their exploits defined the meaning of the activity for generations of Americans.[11]

As middle-class men became devoted hunters, sport hunting won new popularity among the wealthy. The same surge of economic growth that produced an urban middle class also created a new class of monied elites. On the eve of the Civil War, American millionaires numbered a few dozen. By 1900 their ranks had grown to more than 4,000. Status-seeking consumed the lives of the newly wealthy. Lavish parties, opulent dinners, and extravagant balls became hallmarks of industrial fortunes. Sport hunting figured in the recreational repertoire of the upper classes in two ways. On the one hand, it represented a tie to age-old traditions favored by European nobles and other elites. Informed observers had long regarded the sporting pastimes of the English gentry as emblematic of vigorous manhood, and many Americans saw them as worthy of emulation. At the same time, hunting quickly assumed a role in the pastimes that ultimately formed the basis of "country life." With improved transportation systems and the advent of the automobile, wealthy Americans created large estates in pastoral settings near major cities. Long Island became the epicenter of the phenomenon. Between the Civil War and World War II, captains of industry and finance created 975 estates between the city limits of New York and the town of Montauk on the eastern edge of the island. Wealthy Americans also created estates on the outskirts of cities such as Philadelphia and Boston and in the Berkshire Mountains. "Country life," as the associated lifestyle became known, centered on sporting pursuits such as horse racing, yachting, polo, golf, and tennis. Hunting figured among its core activities. As conspicuous consumption became pervasive, hunting remained a privileged part of upper-class culture, simultaneously a tie to longstanding traditions and a continuing source of virtue.[12]

Throughout the last quarter of the nineteenth century, sport hunters enjoyed exceptional conditions. Virtually everywhere they found wildlife in large numbers. In an era of abundance before bag limits, sportsmen killed huge numbers of animals. Even though sporting manuals urged restraint and many sportsmen did the same, hunters nonetheless frequently killed more animals than they could possibly consume. Meanwhile, other hunters also killed large quantities. Not only did sport hunting grow dramatically after the Civil War, commercial hunting also did. Urbanization created unprecedented demand for fresh game. As millions left farms for cities, they retained tastes developed in rural settings. Market hunters killed fowl and other game, loaded it into waiting barrels packed with ice, and sold it to dealers who shipped it by train to major cities.[13]

Although some writers have recalled the late nineteenth century as a "golden age" when "game seemed inexhaustible, seasons were long, [and] bag limits large or nonexistent," it also marked the beginning of fierce struggles over wildlife and undeveloped land. The growth of sport and market hunting sent wildlife populations across the North and Middle West plummeting.

As early as the early 1870s, many hunters voiced alarm at the conditions they saw. As the years passed, the scale of the crisis became fully apparent. Sportsmen immediately took steps to protect their interests. They formed hunting clubs that secured undeveloped land and sought to attract wildlife. Clubs sought to reverse the decline of wildlife populations while ensuring access to existing populations. Their numbers grew dramatically. Sportsmen organized nearly 100 clubs during the winter of 1874–1875 and by 1878, more than 300 existed across the United States, mainly in the Northeast. Sportsmen also lobbied state legislatures to pass laws aimed at protecting wildlife populations. Characterizing subsistence hunters as "pot hunters" and market hunters as "butchers," they portrayed sport as the only legitimate form of hunting. Laws instituting closed seasons and daily and seasonal bag limits resulted. Finally, sportsmen sought out new domains. Traveling to new hunting grounds secured access to wildlife while adding intrigue and adventure. As the popularity of sport hunting grew, travel became an important part of the experiences it provided for many men.[14]

The South figured at the forefront of sportsmen's exploitation of new territory. Amid growing travel to the southern states for tourism, pleasure, and business, sportsmen flocked to the region. In the decades after Reconstruction, middle- and upper-class men from northern cities bought and leased millions of acres in the states of the former Confederacy. Attracted by low land prices, plentiful game, large expanses of undeveloped land, and the image of the South as a veritable Eden, sportsmen seized opportunities to secure access to prime hunting grounds. Old plantation districts proved especially popular. Sportsmen traveled to the coasts of Georgia and South Carolina, northern Florida, south-central Alabama and Mississippi, and southern Louisiana. Upland areas also attracted strong interest. Sportsmen established hunting clubs to secure and manage territory and also established preserves and retreats. They formed relationships with white southerners and employed African American as laborers for tasks ranging from farm labor and carpentry to guiding, driving game, and caring for horses and dogs. In many cases, sportsmen became regular visitors to select areas and important contributors to local economies. As intermittent residents with long-term interests, they assumed important roles.[15]

The overlap between old plantation districts and prime hunting grounds quickly led to the establishment of "hunting plantations." As Scott E. Giltner has recently shown, sportsmen frequently adapted plantations for their use. Plantations supplied much of the basic infrastructure that sportsmen needed: lodgings; barns, sheds, and outbuildings; and fields and forests. Moreover, plantations typically came with a built-in labor force. African American tenants, sharecroppers, and agricultural laborers who lived on or had ties to plantations generally worked for sportsmen in various capacities, whether as laborers, farmhands, guides, or domestics. In some cases sportsmen continued to operate plantations more or less as before, often with modest levels of commercial production. In other cases they limited their efforts to growing produce for consumption onsite. Regardless of the different approaches taken, all adapted plantations for new purposes. Whether sportsmen made substantial changes or minimal alternations, they turned sites formerly used for commercial agriculture into venues for outdoor recreation.[16]

Material development set the sporting plantations of the Red Hills and the lowcountry apart. In each region, sportsmen and sportswomen began developing handsome estates early on. In the Red Hills, the process got underway in the 1880s and 1890s, when wealthy northern families began buying antebellum estates. In 1887, J. T. Metcalfe, a New York doctor, bought Cedar Grove plantation from the Blackshear family, one of the oldest planting families in the Red Hills.[17] In the early 1890s, H. M. Hanna, a director of Standard Oil, bought a plantation once owned by another prominent family and renamed it Pebble Hill.[18] These early purchases set off a boom in land acquisition. In the lowcountry, estates with a strong measure of aesthetic pretention took shape during the 1910s. In Georgetown County, Baltimore pharmaceutical manufacturer Isaac E. Emerson created Arcadia, a 10,000-acre estate composed of lands associated with ten antebellum plantations. Emerson renovated and enlarged the Federal-era mansion at Prospect Hill Plantation, developed extensive gardens, and built a series of new outbuildings. By the early 1910s he owned one of the most elegant estates on the South Atlantic coast. Immediately to the south, financier Bernard M. Baruch amassed a 14,500-acre estate that he named "Hobcaw Barony" after a colonial-era land grant. Baruch and his family used an existing Queen Anne-style dwelling as their residence. On the western branch of the Cooper River, Benjamin and Elizabeth Kittredge purchased Dean Hall Plantation, a handsome estate that they immediately redeveloped as a sporting retreat. Far to the south in Beaufort County, R. H. McCurdy and his wife built a rambling bungalow at Tomotley Plantation, a 5,600-acre tract on the Pocotaglio River.[19]

The creation of large, well-appointed estates reflected far more than facile interest in displaying wealth and status. They made clear owners' intentions of hosting friends, family, and business associates in settings explicitly styled to impress. Upper-class culture placed a premium on socializing amid sumptuous, elegant surroundings. The townhouses of New York's Upper East Side, the country houses of Long Island and Philadelphia's Main Line, urban social clubs, and resort hotels all displayed power and authority. All sought to legitimate wealth. The sporting plantations of the Red Hills and the lowcountry fit this model. Although they varied in scale, form, and styling, all served multiple roles—as private hunting retreats, as seasonal residences, and as venues for hosting persons of comparable social standing. By combining the infrastructure needed for leisure and recreation and material environments that displayed wealth and status, hunting plantations fulfilled practical and symbolic aims.[20]

Although wealthy Americans created similar estates in other southern locales, none came close to matching the sporting plantations of the Red Hills and the lowcountry. The number and character of those in each region proved distinctive. Development did not occur in coordinated fashion. Individually and collectively, the estates of each region hinged upon discrete circumstances and arose largely through the initiative of landowners and prospective buyers. Interaction between owners of estates in the two regions proved limited. Despite common aims, interests, and social and cultural backgrounds, little communication occurred. Not until the 1930s, when widespread fears about wildlife depletion inspired mass action among sportsmen and sportswomen, did significant exchanges take place, and then they centered mainly on concerns about quail habitat and the growing role of naturalist Herbert Stoddard as an expert consultant. Still, similar backgrounds united estate owners and the two regions. Together, they outline circumstances that fall short of explaining the origins of the new estates but nonetheless identify conditions that contributed to their development.[21]

The Red Hills and the lowcountry had flourished and then suffered in ways that made them appealing to would-be estate builders. Plantation agriculture dominated both areas before the Civil War. From the early eighteenth century through to the closing stages of the war, the lowcountry had ranked among the wealthiest plantation districts in the South. Rice, the region's main export, commanded handsome prices on the international market. From the 1740s until the American Revolution, lowcountry planters also produced indigo. About 1800, some turned to sea island cotton, a long-staple cotton that thrived on the coastal islands. Throughout the antebellum era, lowcountry plantations produced strong returns. Geographic constraints and the long-term stability of plantation agriculture fostered development of an exceptionally wealthy and powerful planter class, many of whom owned multiple plantations and lived in a manner reminiscent of European nobles. As Eric Foner has written, "If ever a set of planters seemed to fit the image of the wealthy grandee, it was the rice aristocracy in its golden age." A vast web of business and personal relations, repeated intermarriages, handsome profits, and exceptional social and political influence made lowcountry planters an extraordinary group.[22]

The Civil War sent lowcountry agriculture into a tailspin. Rice and sea island cotton production resumed after the end of hostilities but at sharply reduced levels. Postbellum rice production peaked in 1879 at about a third of the typical annual output of the late antebellum years. The demise of the crop came in the 1890s and first decade of the twentieth century, when a series of devastating storms, increasing competition from domestic producers, and the long-term effects of planters' inability to obtain sufficient labor made it impossible for all but a few to continue. By the early 1910s, only a small number of plantations on the Combahee River continued commercial operations. Sea island cotton remained viable until 1907, when production took a downward turn. Declining crop quality due to hybridization and increasing competition from growers in Florida and the Caribbean hurt South Carolina producers. The arrival of the boll weevil in 1917 sounded the death knell for lowcountry production.[23]

Red Hills plantations produced cotton. Planters settled the area in the 1820s and 1830s and immediately developed large plantations. By the eve of the Civil War, seventy-four encompassed more than 1,000 acres each. Of the region's slaveholders, 119 owned over fifty slaves apiece. While Red Hills planters did not attain the spectacular prosperity that some of their counterparts enjoyed, they nonetheless secured a place for themselves in the upper echelon of southern society.[24]

Emancipation hit Red Hills planters with devastating force. Across the South, sharecropping and tenancy developed in response to freedpeople's desire for autonomy; the difficulty of paying agricultural workers in a cash-poor society; and planters' desire to restart production. In the Red Hills, both arrangements rapidly became the norm. By 1900, tenants operated 56 percent of all farms in Thomas County and 73 percent of those in Leon County. Many planters managed to retain title to their lands, however, and some even expanded their holdings. In fact, the number of large landholdings in the Red Hills grew between 1860 and 1900. Yet landownership offers a misleading picture. Declining prices and weak production left many farmers struggling. By the early 1880s, some resigned to selling out, and others followed in later years. Agriculture in the Red Hills suffered the same downward spiral that befell cotton production in most parts of the South. With international markets well supplied and competition from foreign producers increasing, planters, sharecroppers, and tenants found themselves bound together in a losing struggle, all facing steep odds in a battle to make ends meet.[25]

By the beginning of the twentieth century, the Red Hills and the lowcountry shared a great deal. Both had been home to large plantations before the Civil War; now, decay and decline marked the landscapes of both regions. In the lowcountry, flooded and overgrown rice fields and weathered plantation complexes told a harrowing tale of upheaval and loss. In the Red Hills, patchwork assemblages of sharecropping tracts and tenant farms denoted once-profitable plantations broken up into small parcels. Large tracts of land remained common in both regions. Landowners in the Red Hills and the lowcountry managed to keep large parcels intact, in contrast to the fragmentation that occurred in many areas. Tourism also assumed a presence.

Thomasville, the county seat of Thomas County, thrived as a winter resort from the late 1870s to the mid-1890s, reputedly attracting as many as 25,000 visitors in some years. Luxurious tourist hotels hosted visitors from cities such as New York, Boston, Chicago, and Cleveland, many of them attracted by the supposed health benefits of the region's pine-scented air.[26] The lowcountry did not develop comparable attractions for another several decades, but even in the late nineteenth century small numbers of visitors traveled to see the renowned gardens at Magnolia and Middleton Place plantations and to winter at Summerville's Pine Forest Inn. The growing popularity of Florida resorts brought additional visitors to the region and eventually prompted business and civic leaders to launch efforts aimed at capturing the "tourist trade."[27]

In short, similar patterns of development gave the Red Hills and the lowcountry a set of shared attributes that suggest the influence of factors beyond inexpensive land and plentiful wildlife. Exactly what encouraged the creation of plantations devoted to upper-class leisure and recreation remains unclear, however. The essays in this collection take steps toward a fuller understanding of these and other problems. They bring needed attention to questions that have long been apparent but have heretofore evaded sustained investigation.

The essays that follow are loosely organized around three major themes. Collectively, they provide a framework for investigating the new estates and identifying problems associated with their development. *Social relations* considers the relationships that developed between sportsmen and sportswomen and white and black southerners. The essays concerned with this topic explore the crucial roles that native southerners played in the creation and upkeep of the new estates, sportsmen's and sportswomen's interactions with African Americans, and southerners' views of the newcomers and their plantations. Ultimately, they foreground the centrality of race, class, and sectional identities in the making of northerners' plantations. *Environmental change* considers how sporting plantations affected the landscape and ecology of the Red Hills and the lowcountry. Questions about land use, management practices, and the interplay of natural systems and human activity occupy this category. Although historians have tended to characterize land and wildlife conservation as "passive" activities that allowed land and game populations to revert back toward a "natural" state, in fact, each imposed new demands. Wealthy sporting enthusiasts instituted practices aimed at encouraging reproduction of select species, development of game habitat, and exclusion of subsistence hunters from protected areas. Understanding these changes illuminates the environmental history of the southeastern United States and the history of land and wildlife conservation more generally. It also highlights the scale of northerners' activities and their influence on traditional forms of land use. *Culture* considers the meaning and symbolism of plantations remade for leisure. It considers what people meant by the terms "estate" and "plantation," how onlookers viewed the new estates, and how patterns of physical change and activity influenced perceptions of various actors. It also considers the experiences of sportsmen and sportswomen and their guests.

A fourth theme that runs through several essays concerns spatial and material attributes. Drawing inspiration from studies of landscape, architecture, and place, essays by Jennifer Betsworth, Drew Swanson, Hayden Ross Smith, and Robin Bauer Kilgo pay close attention to the material form and spatial organization of sporting estates and patterns of material change. Collectively, these studies highlight shifting conceptions of plantations and their physical manifestation. As historical uses ended and new purposes developed, design and material form took on new significance. Indeed, by the 1920s, some observers viewed material features as definitive, at least as much so as historical uses. In creating their estates, northerners retained select features from earlier periods, destroyed others, and added new buildings and landscaping in what appeared to most observers as an evolutionary process, respectful of the past but not beholden to it. In studying these developments, Betsworth, Swanson, Smith, and Bauer Kilgo pay close attention to the changes that occurred and what they sought to achieve. They also reveal intense concerns for place. What arises from these essays is a view of plantations that foregrounds material conditions and modes of sensory experience over other qualities. Whereas historians have generally seen the combination of staple agriculture and unfree labor as definitive, the essays in this collection show a different view. As agriculture ceased to be remunerative in some parts of the South, people increasingly saw plantations in other ways. These essays illuminate this process and its consequences.[28]

All of the essays examine plantations during a moment of transformative change. Several questions thread through them all. What did plantations remade for leisure become? What did they offer to owners and others, and how did contemporaries understand them? The rise of large sporting estates in the Red Hills and the lowcountry occurred swiftly, within about a generation. When viewed in a historical perspective, it occurred with dizzying speed. In the span of roughly four decades, plantations went from being distressed sites of export-oriented agriculture to handsome estates. How people reacted and saw the changes underway stand among the questions that animate the essays in this volume.

The investigation of culture and social relations begins with chapters by Daniel Vivian, Jennifer Betsworth, and Drew Swanson. Vivian's essay, "'Plantation Life': Varieties of Experience on the Remade Plantations of the South Carolina Lowcountry," tackles questions about northerners' activities through sustained analysis of what became known as "plantation life." Challenging the conventional view that mild weather, plentiful wildlife, and the lingering mystique of once-grand plantations explains the rise of the new estates, Vivian examines the activities that brought northerners to the lowcountry and the experiences they afforded. Emphasizing social and environmental conditions, he finds that northerners viewed plantations as distinctively southern spaces where the exoticism, racial hierarchies, and decaying grandeur popularly associated with "the South" seemed especially pronounced. Northerners viewed African American laborers as innate to plantations and hunting in the South as emulating the supposedly stable hierarchies of the antebellum era. The style of hunting that northerners' practiced had roots in aristocratic traditions and close ties to country life. Northerners did not come simply to enjoy existing conditions; they shaped their estates to meet deeply felt needs. By appropriating select dimensions of "the South" and associating them with contemporary practices and beliefs, northerners created an experiential realm that affirmed their sense of class privilege while offering invigorating, challenging experiences and opportunities for virtuous endeavor. In short, plantation life proved uniquely suited to heirs of great wealth and industrial and financial moguls in an era of upper-class decline.

In "Restoring and Reviving Southern Ruins: Reshaping Plantation Architecture and Landscapes in Georgetown County, South Carolina," Betsworth examines the changing use and purpose of plantations from the perspective of material form and aesthetics. In studying the choices northerners made concerning architecture and landscape design, Betsworth finds strong favor for restoring early houses, building Colonial Revival mansions, and creating landscapes that appeared centuries old. Northerners spared little effort and expense in creating estates centered upon stylish houses, groomed lawns, and elaborate gardens. In the process, they turned well-known plantations into sites of grandeur and sumptuous display. At the same time, they obscured historical evidence of slavery. In exploring these processes, Betsworth demonstrates the scale of change that occurred and highlights its most important consequences. Ultimately, she compels reexamination of traditional characterizations of northerners' efforts and suggests ample opportunities for further inquiry.

Drew Swanson brings a somewhat different pattern of change to light in investigating the making of Wormsloe Gardens, a tourist attraction created at Wormsloe Plantation, a site that historically produced sea island cotton near Savannah. The sole essay to focus on a plantation on the Georgia coast, Swanson's study reminds us that the remaking of old plantations for new uses occurred outside of the Carolina lowcountry and the Red Hills. Most examples in coastal Georgia seized upon the combination of mild weather, spectacular scenery, and outdoor recreation that drew wealthy sportsmen and sportswomen to other parts of the South. At Wormsloe, historical memory, landscape design, and commercial opportunities combined to produce a tourist attraction similar to the better-known plantation gardens at Middleton Place and Magnolia plantations near Charleston. Wymberly W. and Augusta De Renee, fifth-generation descendants of Wormsloe's original master, Noble Jones, acted out of deep commitment to their family legacy and an abiding love of their ancestral seat in creating gardens and a set of associated attractions that they hoped would prove popular enough to extract them from debt. Although the effort failed to achieve their immediate goal, it navigated broad shifts in the southern economy by turning a once-productive enterprise into a site of leisure and consumption. In charting this transition, Swanson highlights development of historical and horticultural attractions and the erasure that occurred as interpretation of Wormsloe's history focused on an idealized past.

Patterns of landscape change and environmental history figure at the center of Matthew Lockhart's essay, "'Rice Planters in their Own Right': Northern Sportsmen and Waterfowl Management on the Santee River Plantations During the Baiting Era, 1905–1935." In tracing the efforts of the Santee Gun Club to ensure high-quality shooting for its members, Lockhart shows that large-scale rice planting did not end with the collapse of commercial production but, rather, continued for another several decades on an impressive scale. The Santee Club continued planting in order to attract ducks to its lands, a practice that other clubs and owners of sporting estates also adopted, albeit less intensively. Lockhart shows that sportsmen made astounding efforts in pursuit of excellent shooting conditions. His essay also overturns the conventional wisdom about the end of rice cultivation on the Carolina coast while calling attention to the history of private efforts at waterfowl management. While most histories locate the beginnings of conservation-inspired manipulation of wetlands to the development and expansion of national wildlife refuges during the New Deal, Lockhart shows that duck-hunting clubs started decades earlier. Although the extent of similar efforts is unknown, the example of the Santee Club is nonetheless significant.

Questions about access to wildlife and the intersections of landscape, power, and recreation also figure in chapters by Hayden Ross Smith and Julia Brock. Smith's essay, "Knowledge of the Hunt: African American Guides in the South Carolina Lowcountry at the Turn of the Twentieth Century," examines knowledge of lowcountry ecosystems that African Americans developed under slavery and continued to cultivate after emancipation. Through detailed investigation of original sources, Smith finds that African Americans developed exceptional knowledge of the lowcountry landscape and animal behavior because of tasks assigned by planters and planters' encouragement of hunting for subsistence. African Americans cultivated mastery of hunting because it offered status, autonomy, and opportunities to escape routine labor. After the Civil War, sportsmen prized black guides for their expert knowledge. Guiding became a means of earning income and status. In showing how African American hunting guides seized opportunities that arrived with northern sportsmen, Smith reveals the ironies of power and authority in the rural lowcountry of the early twentieth century. Knowledge and skills that planters fostered under slavery became a source of power afterward and created unique opportunities for former slaves and their descendants.

Brock's investigation explores the relationships of class and power that developed in southern Georgia with the creation of large sporting estates. In "A 'Sporting Fraternity': Northern Hunters and the Transformation of Southern Game Law in the Red Hills Region, 1880–1920," she finds that class alliances between northern sportsmen and local elites proved crucial to the development of large estates. The same relationships also undergirded efforts to limit access to wildlife. Large landowners lobbied for adoption of new game laws that limited the customary rights of landless whites and blacks and small farmers, which resulted in an outcry from the latter. Brock's essay highlights the social and political circumstances that surrounded the development of sporting estates and how northerners' activities influenced local and regional power relations. Ultimately, she shows that the creation of Red Hills estates had greater consequences than commonly recognized.

The final essay, Robin Bauer Kilgo's "Life and Labor on the Southern Sporting Plantation: African American Tenants at Tall Timbers Planation, 1920–1944," shifts the focus to landless farmers who lived and labored on sporting estates. Drawing on an exceptional body of sources, Kilgo examines the conditions of material life and social relations at Tall Timbers Plantation in northern Leon County, Florida. Bauer finds that tenants enjoyed somewhat better conditions than their counterparts on agricultural plantations but still experienced manifold hardships and had few opportunities. Tenancy continued at Tall Timbers throughout the New Deal, in part because of accommodations made by the owner, Henry Beadel, but it collapsed rapidly during the 1940s. Tenants seized opportunities for military service, defense-industry jobs, and other forms of employment in nearby cities. Tenancy thus met a similar fate as it did on plantations across the South, despite Beadel's leniency.

Although the essays in this collection explore diverse territory, they leave no doubt about the significance of sporting estates and the need for further investigation. Neither old plantations repurposed for leisure nor estates called "plantations" simply for the sake of perceived grandeur, the sporting plantations of the Red Hills and lowcountry occupied terrain all their own. Their history underscores the complexity of land, labor, and leisure in the post-emancipation South and the importance of shifting patterns of North-South interaction. In exploring select dimensions of their development, this collection takes steps toward critical investigation of long-overlooked histories. The results include new perspectives on the South's role in American culture and the history of two of its most distinctive subregions.

NOTES

1. "From Thomasville to Tallahassee," *Country Life* 67, no. 4 (Feb. 1935): p. 11.
2. James C. Derieux, "The Renaissance of the Plantation," *Country Life* 41, no. 3 (Jan. 1932): pp. 34–39.
3. See, for example, Derieux, "The Renaissance of the Plantation"; "Charleston, South Carolina," *Fortune* 7, no. 3 (Mar. 1933): pp. 78–81, 83; "South Carolina Plantations," *Town and Country* 90, no. 4144 (Jan. 15, 1935): pp. 31–33; "The Low Country," *Life* 7, no. 26 (Dec. 25, 1939): pp. 38–45; "From Tallahassee to Thomasville."
4. "From Thomasville to Tallahassee"; "The Renaissance of the Plantation."
5. George Brown Tindall, *The Emergence of the New South, 1913–1945* (Baton Rouge: Louisiana State University Press, 1967), p. 103; William R. Brueckheimer, "The Quail Plantations of the Thomasville-Tallahassee-Albany Regions," *Journal of Southwest Georgia History* 3 (fall 1965): pp. 44–63; Clifton Paisley, *From Cotton to Quail: An Agricultural Chronicle of Leon County, Florida, 1860–1967* (Gainesville: University of Florida Press, 1968), pp. 74–98; George C. Rogers, Jr., *The History of Georgetown County, South Carolina* (Columbia: University of South Carolina Press, 1970), chap. 21; William Warren Rogers, *Thomas County, 1865–1900* (Tallahassee: Florida State University Press, 1973); William Warren Rogers, *Transition to The Twentieth Century: Thomas County, Georgia, 1900–1920* (Tallahassee: Sentry Press, 2002); William Warren Rogers, *Pebble Hill: The Story of a Plantation* (Tallahassee: Sentry Press, 1979); William Warren Rogers, *Foshalee: Quail Country Plantation* (Tallahassee: Sentry Press, 1989).
6. Albert G. Way, *Conserving Southern Longleaf: Herbert Stoddard and the Rise of Ecological Land Management* (Athens: University of Georgia Press, 2011); Hank Margeson and Joseph Kitchens, *Quail Plantations of South Georgia and North Florida* (Athens: University of Georgia Press, 1991); Stuart A. Marks, *Hunting in Black and White: Nature, History, and Ritual in a Carolina Community* (Princeton: Princeton University Press, 1990); Mark Wetherington, *The New South Comes to Wiregrass Georgia, 1860–1910* (Knoxville: University of Tennessee Press, 1994); Peter A. Coclanis, *The Shadow of a Dream: Economic Life and Death in the South Carolina Low Country, 1670–1920* (New York: Oxford University Press, 1989), p. 156; Stephanie E. Yuhl, *A Golden Haze of Memory: The Making of Historic Charleston* (Chapel Hill: University of North Carolina Press, 2005), pp. 177–183; Virginia Beach, *Medway* (Charleston: Wyrick and Co., 1999); David De Long, *Auldbrass: Frank Lloyd Wright's Southern Plantation* (New York: Rizzoli, 2003); Robert B. Cuthbert and Stephen G. Hoffius, *Northern Money, Southern Land: The Lowcountry Plantation Sketches of Chlotilde R. Martin* (Columbia: University of South Carolina Press, 2009); Lee G. Brockington, *Plantation Between the Waters: A Brief History of Hobcaw Barony* (Charleston: History Press, 2009), chaps. 4–6; Larry R. Youngs, "Lifestyle Enclaves: Winter Resorts in the South Atlantic States, 1870–1930 (Florida, Georgia, South Carolina, North Carolina)" (Ph.D. diss., Georgia State University, 2001).
7. "From Thomasville to Tallahassee," p. 13. For other discussions of estate owners' activities, see Chalmers S. Murray, "Attracted by Climate Northerners Become

Land Barons of Carolina Coast," *News and Courier* (Charleston, S.C.), Dec. 15, 1929, p. A3; Chlotilde R. Martin, "Low-Country Plantations Stir as Air Presages Coming of New Season," *News and Courier*, Oct. 16, 1932, p. A6; "South Carolina Plantations," p. 31; Edward A. Lowry, "Winter Colonies in the South," *Munsey's Magazine* 28, no. 3 (Dec. 1902): pp. 325–333; "The Growth of Thomasville," *Southern World* (Atlanta, Ga.), Nov. 15, 1883, p. 35; "Many Refuges from the Cold–The Country Full of Healthy Winter Resorts," *New York Times* (New York, N.Y.), Jan. 3, 1892, p. 20; "With Gun and Dog in Georgia," *Forest and Stream*, Jan. 10, 1903, p. 29; H. C. S., "A Georgia Quail Country," *Forest and Stream*, Feb. 24, 1893, p. 161. For laments about social obligations, see, for example, Sydney J. Legendre, "Diary of Life at Medway Plantation, Mt. Holly, South Carolina [1937–1941]," entries for May 9, 1939, and Mar. 29, 1940 (private collection, copy in possession of Daniel Vivian); Thomas A. Stoney Journal, entries for Feb. 12 and Mar. 17, 1937, in Boone Hall Scrapbook No. 3, South Carolina Historical Society, Charleston, S.C. (hereafter SCHS).

8. See, for example, Martin, "Low-Country Plantations Stir as Air Presages Coming of New Season"; Susan Lowndes Allston, "Windsor Place Spot of Beauty," *News and Courier*, Dec. 22, 1929, p. A3; "Big Sums Spent on Plantations," *News and Courier*, Nov. 8, 1936, p. C3.

9. On the history of plantations and their relationship to racial slavery, see especially Philip D. Curtin, *The Rise and Fall of the Plantation Complex: Essays in Atlantic History*, 2nd ed. (New York: Cambridge University Press, 1990); Robin Blackburn, *The Making of New World Slavery: From the Baroque to the Modern, 1492–1800* (New York: Verso, 1997). On the problems associated with sharecropping and tenancy during the early twentieth century, see especially Frank Tannenbaum, *Darker Phases of the South* (New York: G. P. Putnam's Sons, 1924), chap. 4; Arthur F. Raper, *Preface to Peasantry: A Tale of Two Black Belt Counties* (Chapel Hill: University of North Carolina Press, 1936); Pete Daniel, *Breaking the Land: The Transformation of Cotton, Tobacco, and Rice Cultures Since 1880* (Urbana, Ill: University of Illinois Press, 1985), p. 240; Natalie J. Ring, *The Problem South: Region, Empire, and the New Liberal State, 1880–1930* (Athens: University of Georgia Press, 2012), chap. 3.

10. Daniel Justin Herman, *Hunting and the American Imagination* (Washington, D.C.: Smithsonian Institution Press, 2001), pp. 128–172, 188–199; David Strauss, "Toward a Consumer Culture: 'Adirondack Murray' and the Wilderness Vacation," *American Quarterly* 39, no. 2 (summer 1987): pp. 270–286; David E. Shi, *The Simple Life: Plain Living and High Thinking in American Culture* (New York: Oxford University Press, 1985), pp. 155, 175–214.

11. Herman, *Hunting and the American Imagination*, chap. 16. On overwork and its consequences, see E. Anthony Rotundo, *American Manhood: Transformations in Masculinity from the Revolution to the Modern Era* (New York: Basic Books, 1993), pp. 185–193; T. J. Jackson Lears, *No Place of Grace: Antimodernism and the Transformation of American Culture, 1880–1920* (New York: Parthenon Books, 1981), pp. 4–58. On shifting terrain of American manhood during the era, see especially Michael Kimmel, *Manhood in America: A Cultural History*, 3rd ed. (New York: Oxford University Press, 2012), chaps. 3–5; Gail Bederman,

Manliness and Civilization: A Cultural History of Gender and Race in the United States, 1880–1917 (Chicago: University of Chicago Press, 1995). On the relationship between sport, masculinity, and religion, see Clifford Putney, *Muscular Christianity: Manhood and Sports in Protestant America,* 1880–1920 (Cambridge, Mass.: Harvard University Press, 2001).

12. Sven Beckert, *The Monied Metropolis: New York City and the Consolidation of the American Bourgeois,* 1850–1896 (New York: Cambridge University Press, 2001), chaps. 5 and 8; Eric Homberger, *Mrs. Astor's New York: Money and Social Power in a Gilded Age* (New Haven: Yale University Press, 2002); Frederic Cople Jaher, "The Gilded Elite: American Multimillionaires, 1865 to the Present," in *Wealth and the Wealthy in the Modern World,* ed. W. D. Rubenstein (New York: St. Martin's Press, 1980): pp. 190–201; Mary Cable, *Top Drawer: American High Society from the Gilded Age to the Roaring Twenties* (New York: Atheneum, 1984); Shi, *The Simple Life,* pp. 154–164. On country estates and associated activities, see Clive Aslet, *The American Country House* (New Haven: Yale University Press, 1990); Mark Alan Hewett, *The Architect and the American Country House,* 1890–1940 (New Haven: Yale University Press, 1990); Robert B. MacKay, Anthony K. Baker, and Carol A. Traynor, *Long Island Country Houses and their Architects,* 1860–1940 (New York: Society for the Preservation of Long Island Antiquities in association with W. W. Norton and Co., 1997); Richard S. Jackson and Cornelia B. Gilder, *Houses of the Berkshires:* 1870–1930 (New York: Acanthus Press, 2006); William Morrison, *The Main Line: Country Houses of Philadelphia's Storied Suburb,* 1870–1930 (New York: Acanthus Press, 2002).

13. Herman, *Hunting and the American Imagination,* pp. 204–205, 237–253; James A. Tober, *Who Owns the Wildlife?: The Political Economy of Game Conservation in Nineteenth-Century America* (Westport, Conn.: Greenwood Press, 1981), chaps. 1–3.

14. Bob Hinman, *The Golden Age of Shotgunning* (New York: Winchester Press, 1971), p. 2; Tober, *Who Owns the Wildlife,* pp. 43, 49; Herman, *Hunting and the American Imagination,* pp. 237–253.

15. Giltner, *Hunting and Fishing in the New South.* On postbellum tourism in general, see Nina Silber, *The Romance of Reunion: Northerners and the South,* 1865–1900 (Chapel Hill: University of North Carolina Press, 1993), chap. 3; W. Fitzhugh Brundage, *The Southern Past: A Clash of Race and Memory* (Cambridge, Mass.: Belknap Press of Harvard University Press, 2005), chap. 5; Anthony J. Stanois, ed., *Dixie Emporium: Tourism, Foodways, and Consumer Culture in the American South* (Athens: University of Georgia Press, 2008); Richard D. Starnes, ed., *Southern Journeys: Tourism, History, and Culture in the Modern South* (Tuscaloosa: University of Alabama Press, 2003); Karen L. Cox, *Destination Dixie: Tourism and Southern History* (Gainesville: University Press of Florida, 2012).

16. Giltner, *Hunting and Fishing in the New South,* chap. 4.

17. Brueckheimer, "The Quail Plantations of the Thomasville-Tallahassee-Albany Regions," p. 54.

18. Ibid., pp. 55–56.

19. Chalmers S. Murray, "Arcadia, Where LaFayette Stopped," *News and Courier,* June 21, 1931, p. B5; Brockington, *Plantation Between the Waters,* pp. 40, 49–55;

Chlotilde R. Martin, "The Cypress Gardens, In Berkeley," undated newspaper clipping from *News and Courier* in Dean Hall Plantation vertical file, SCHS; John B. Irving, *A Day on Cooper River*, 2nd ed., ed. Louisa Cheves Stoney (Columbia: R. L. Bryan Co., 1932), pp. 28–29; Chlotilde R. Martin, "Trees Hang Over Way to Tomotley," *News and Courier*, Nov. 23, 1930, p. A12; Suzanne Cameron Linder and Marta Leslie Thacker, *Historical Atlas of the Rice Plantations of Georgetown County and the Santee River* (Columbia: South Carolina Department of Archives and History for the Historic Ricefields Association, Inc., 2001), pp. 53–54, 77.

20. Wayne Craven, *Gilded Mansions: Grand Architecture and High Society* (New York: W. W. Norton and Co., 2009); Aslet, *The American Country House*; Hewett, *The Architect and the American Country House*; MacKay et al., *Long Island Country Houses*; Cable, *Top Drawer*, pp. 51–58, 85–113, 135–159.

21. For an overview of estate development and the rise of other recreational venues in the Southeast, see Drew A. Swanson, *Remaking Wormsloe Plantation: The Environmental History of a Lowcountry Landscape* (Athens: University of Georgia Press, 2012), pp. 142–146. On Stoddard's consultancy, see Way, *Conserving Southern Longleaf*, chap. 5; Herbert Stoddard, *The Bobwhite Quail: Its Habits, Preservation, and Increase* (New York: Charles Scribner and Sons, 1931); Stoddard, *Memoirs of a Naturalist* (Norman: University of Oklahoma Press, 1969). On lowcountry estate owners' interaction with Stoddard and associates, see, for example, Legendre, "Diary of Life at Medway Plantation," entries for May 9, 1939, and Feb. 26, 1940; and correspondence in folder labeled "Cain Hoy Plantation. Quail, 1935-'41," box 133, Harry F. Guggenheim Papers, Library of Congress, Washington, D.C.

22. Eric Foner, *Nothing But Freedom: Emancipation and its Legacy* (Baton Rouge: Louisiana State University Press, 1983), pp. 77; *Walter Edgar, South Carolina: A History* (Columbia: University of South Carolina Press, 1998), pp. 139–154, 161–163, 189–199, 265–270; Peter H. Wood, *Black Majority: Negroes in Colonial South Carolina from 1670 through the Stono Rebellion* (New York: Knopf, 1974); S. Max Edelson, *Plantation Enterprise in Colonial South Carolina* (Cambridge, Mass.: Harvard University Press, 2006); Joyce E. Chaplin, *An Anxious Pursuit: Agricultural Innovation and Modernity in the Lower South, 1730–1815* (Chapel Hill: Institute of Early American History and Culture by University of North Carolina Press, 1993); Philip D. Morgan, *Slave Counterpoint: Black Culture in the Eighteenth-Century Chesapeake and Lowcountry* (Chapel Hill: Omohundro Institute of Early American History and Culture by University of North Carolina Press, 1998); Robert M. Weir, *Colonial South Carolina: A History* (Columbia: University of South Carolina Press, 1983).

23. Coclanis, *The Shadow of a Dream*, pp. 140–142; James H. Tuten, *Lowcountry Time and Tide: The Fall of the South Carolina Rice Kingdom* (Columbia: University of South Carolina Press, 2010), chaps. 2 and 3; Richard D. Porcher and Sarah Fick, *The Story of Sea Island Cotton* (Charleston: Wyrick and Company, 2005), pp. 327–330; Charles F. Kovacik and Robert E. Mason, "Changes in the South Carolina Sea Island Cotton Industry," *Southeastern Geographer* 25, no. 2 (Nov. 1985): pp. 94–97.

24. Brueckheimer, "The Quail Plantations of the Thomasville-Tallahassee-Albany Regions," pp. 47–50; Way, *Conserving Southern Longleaf*, p. 26.

25. Brueckheimer, "The Quail Plantations of the Thomasville-Tallahassee-Albany Regions," pp. 49–52; Bureau of the Census, *Thirteenth Census of the United States Taken in the Year 1910*, vol. VI, *Agriculture, 1909 and 1910* (Washington, D.C.: Government Printing Office, 1913), pp. 303, 356; Clay Ouzts, "Landlords and Tenants: Sharecropping and the Cotton Culture in Leon County, Florida, 1865–1885," *Florida Historical Quarterly*, 75, no. 1 (summer 1996): pp. 1–23.

26. Way, *Conserving Southern Longleaf*, pp. 27–34; Julia Brock, "Land, Labor, and Leisure: Northern Tourism in the Red Hills Region, 1890–1950" (Ph.D. diss., University of California at Santa Barbara, 2012), chap. 1.

27. John Hammond Moore, comp. and ed., *South Carolina in the 1880s: A Gazetteer* (Orangeburg, S.C.: Sandlapper Publishing, 1989), pp. 258–269; "At the Palace in the Pines," *News and Courier*, Dec. 14, 1908, p. 3; Walter J. Fraser, Jr., *Charleston! Charleston!: The History of a Southern City* (Columbia: University of South Carolina Press, 1989), pp. 368–375.

28. For especially influential studies of space, landscape, and place, see especially Barbara Bender, ed., *Landscape: Politics and Perspectives* (Providence, R.I.: Berg Publishers, 1993); Yi-Fu Tuan, *Space and Place: The Perspective of Experience* (Minneapolis: University of Minnesota Press, 1977); Chris Wilson and Paul E. Groth, eds., *Everyday America: Cultural Landscape Studies after J. B. Jackson* (Berkeley: University of California Press, 2003); Michael Parker Pearson and Colin Richards, eds., *Architecture and Order: Approaches to Social Space* (New York: Routledge, 1994); Allan Pred, "Place as Historically Contingent Process: Structuration and the Time-Geography of Becoming Places," *Annals of the Association of American Geographers* 74, no. 2 (June 1984): pp. 279–297; Barney Warf and Santa Arias, eds., *The Spatial Turn: Interdisciplinary Perspectives* (New York: Routledge, 2009).

Chapter 1

"Plantation Life"

Varieties of Experience on the Remade Plantations of the South Carolina Lowcountry

Daniel Vivian

During the first four decades of the twentieth century, the coastal region of South Carolina became a seasonal playground for some of the wealthiest Americans of the era. At hunting clubs set in secluded locations, men from northern cities socialized with friends and acquaintances while enjoying "good sport." At smaller retreats and preserves, men with similar backgrounds spent their time in roughly comparable fashion. At large estates, wealthy sportsmen and sportswomen entertained, socialized, and engaged in a variety of leisure and sporting pursuits. As transportation improvements and surging industrial wealth placed the South Atlantic coast within easy reach of northern elites, the Carolina lowcountry attracted widespread attention. By the early 1930s a patchwork landscape of private hunting domains stretched throughout the roughly 130 miles between Georgetown and the Savannah River and as far west as Williamsburg County.[1]

The estates that northerners created during the era showed ambitions beyond sport alone. Elegant architecture, elaborate landscaping, and plentiful labor signified their role as venues for socializing, entertaining, recreation, and ritualized performances. Most featured large acreages, landscaped grounds and gardens, and handsome buildings. Owners typically restored and rehabilitated colonial-era and early nineteenth-century houses or built new Colonial Revival mansions. A few opted for more modest dwellings, and some chose different architectural styles. Variations notwithstanding, all of the new estates served as sites of upper-class leisure and recreation. All offered opportunities for enjoying mild weather, peace and quiet, and excellent shooting. In the era when the excesses of the 1920s gave way to the hardships of the Great Depression, wealthy sportsmen and sportswomen made the Carolina coast a premier destination for their favored pursuits.[2]

The "shooting plantations" of the Carolina lowcountry are well known to historians and lay audiences alike. For decades, they have been recalled as part of the "Second Yankee Invasion," a land-buying and estate-making spree that transformed large swaths of the rural lowcountry between the 1890s and World War II. Sportsmen and sportswomen from outside the region—variously called "northerners," "Yankees," and "winter colonists" by natives—created more than seventy estates with large acreages, well-developed building complexes, and coordinated architecture and landscaping. Few survive today. Most, long ago, found themselves turned into residential housing tracts, golf course communities, or wildlife preserves. Still, they continue to anchor remembrance of the era when "rich Yankees" showed new interest in the lowcountry. As symbols of an era when the region gained new life after decades of decline, they occupy a prominent place in regional lore. They also cast light on the origins of the lowcountry that exists today. Since the 1970s, the lowcountry has been synonymous with stunning scenery, outdoor recreation, and historical charm. Beachfront resorts, golf courses, museums, and historic sites attract visitors from across the country and around the world. The sporting estates of the 1920s and 1930s played a crucial role in bringing attention to the region and establishing it as a premier destination for outdoor recreation. Their history forms part of the lowcountry's transition from economic and cultural backwater to tourist mecca—and helps account for the spectacular success of its leisure and tourism industries.[3]

Despite the significance of the new estates, historians have shown them limited attention. Save for passing mentions in a handful of books and articles, they remain largely unexplored. One way to bring their history into focus lies in examination of what contemporaries called "plantation life." By the late 1920s, owners and onlookers used this phrase in referring to the activities that owners and their guests enjoyed during their winter sojourns. As shorthand for the experiences that the new "plantations" afforded, the term signified what plantations across the region had become. As an index of the changes that had occurred, it told of the purposes plantations now served. More than a referent to activities carried on at places historically called plantations, plantation life identified what most people regarded as an evolutionary process. It told more about what the new owners and their guests had created than what plantations had historically been.[4]

Investigating the meaning of "plantation life" brings a complex nexus of behaviors and beliefs into view. Social relations, recreational practices, and environmental conditions informed understandings of northerners' activities. The cultural significance of owners' preferred forms of recreation and attendant social and racial hierarchies also influenced onlookers' perspectives. When people spoke of "plantation life," they referred to a distinctive realm of experience. Not simply a product of old plantations repurposed for recreation, it told of how Americans of the era thought about plantations and their potential. Plantation life merged traditional upper-class sporting pursuits with the social hierarchies and environmental conditions of the rural low-country. It made plantations anew while echoing practices that most observers saw as characteristic of plantations of earlier periods and upper-class life in general. Sportsmen and sportswomen found it eminently satisfying, for it connoted status while supplying encounters with exoticism and moments of vital experience. For members of a social and economic elite living amid a democratizing age, it reassured, affirmed, and sanctioned the value of long-standing traditions.[5]

This essay examines plantation life. It considers the activities that north-erners and their guests engaged in, the social and cultural significance of those activities, and the experiences they provided. It also explores the social and environmental character of northern-owned plantations. The perspective employed mirrors the views of estate owners, their guests, and, to a lesser extent, interested onlookers, mainly white lowcountry elites and journalists. The focus centers on people who used the phrase "plantation life" and who saw northerners' estates as offering desirable varieties of experience. African Americans and small numbers of working-class whites also experienced plan-tation life, but mainly as a form of labor, not leisure. Consequently, it held different meanings for them. Exactly how they viewed it is uncertain, for lim-ited sources make recovering their perspectives difficult. By contrast, the his-torical record offers copious information about estate owners and their guests. For them, plantation life merged the imagined South of popular lore with the physical space and lived experience of plantations adapted for leisured use. It created a performative realm wherein race, southernness, and class privilege continually informed northerners' individual and collective senses of self and mediated their relations with lowcountry people. Ultimately, plantation life became a mark of distinction. Emphasis on virtuous leisure and vigorous rec-reation affirmed northerners' view of themselves while demonstrating status. By assuaging class-based anxieties in an age of social upheaval, plantation life proved crucial to maintenance of class boundaries. It reified individual and collective identities and placed estate owners and guests in coveted roles.

Recreational hunting formed the cornerstone of plantation life. In the 1880s and 1890s, northern sportsmen began traveling to the lowcountry in search of untapped hunting domains. What they found astonished them. With agri-culture in decline, large tracts of land could be bought and leased at low prices. Wildlife abounded. A landscape of coastal swamps, dense forests, and flooded rice fields provided ample feed and cover for large numbers of animals. With wildlife populations closer to home decimated by overhunting, sportsmen moved quickly to take advantage of these conditions. They formed hunting clubs that seized control of tens of thousands of acres and purchased and leased land individually, in pairs, and in small groups. By about 1910, sportsmen from northern cities owned or had secured access to upward of 200,000 acres of coastal lands. In later years, land acquisition continued at a brisk pace.[6]

In the late 1920s, as northerners' activities reached their peak, hunting remained central to the pursuits they came to the lowcountry to enjoy. Conditions had changed over time, and the relative importance of hunting for owners of large estates had also. Game populations had declined from the seemingly inexhaustible levels of earlier years. Legally mandated bag limits and game seasons curtailed heavy game-taking. Yet even with these changes, the lowcountry continued to offer favorable conditions. Sportsmen and sportswomen recognized the region as a prime destination for ducks, deer, wild turkey, and a host of other species. As wildlife became increasingly scarce across the North and the Middle West, the lowcountry held its own.[7]

The style and manner in which sportsmen and sportswomen hunted changed with the growth and development of the lowcountry sporting scene. As the number of large estates proliferated and entertaining friends, family, and business associates became common, hunting became less the focus of sportsmen's and sportswomen's activities than a crucial part of an expanded array of leisure and sporting pursuits. By the late 1920s, most owners also enjoyed horseback riding, carriage rides, fishing, boating, and sightseeing. These activities provided variety, accommodated differing interests, and took advantage of local conditions. By reducing the intensity of hunting, they aided conservation efforts. They also made allowances for novice hunters. As hosting guests became routine, accommodating less-experienced sporting enthusiasts became important. Different forms of recreation and moderately-challenging forms of hunting ensured the participation of persons with limited skills.

Northerners practiced a ritualized style of hunting that emphasized displays of status and authority. Heirs to the sporting traditions of the English gentry and Teddy Roosevelt's vision of "the strenuous life," sportsmen and sportswomen took to the field for more than mere pleasure and enjoyment. Like other Americans of the era, they saw hunting as a vital means of developing and maintaining strength, stamina, skill, and self-control. Hunting afforded encounters with unspoiled nature while at the same time demanding discipline and nerve. Adhering to sporting practices meant passing on shots that would have almost certainly produced a kill. It also meant relishing the experience of hunting no matter what the outcome. The patience required fostered appreciation of nature and the challenges inherent in stalking wildlife. Sport hunting demanded that hunters see the value in foregoing readily attained forms of gratification in favor of success earned through discipline, perseverance, and good fortune. The role of sport hunting in American culture reflected such values. At the beginning of the twentieth century, the title of sportsmen remained a marker of trust and integrity. Biographical profiles of prominent businessmen and politicians identified many as sportsmen not simply for the sake of mentioning a hobby but as indicators of courage, respectability, and probity. At a time when the workings of government and business seemed increasingly opaque, the status of "sportsman" carried significant weight. It recalled an era of face-to-face relationships and personal bonds that to many Americans seemed to have all but disappeared with the rise of large corporations and bureaucratic organizations.[8]

Economic considerations reinforced the cultural significance of sport hunting. An elite activity by nature, sport hunting identified members of a privileged class. The ability to hunt for pleasure demonstrated economic independence. When middle- and upper-class hunters took to the field, they risked coming home empty handed. Sporting practices, after all, made killing wild animals more difficult, not less. By contrast, men who hunted to feed themselves or sell game at market could ill afford to pass up easy kills. They shot animals whenever they could. Beyond these considerations, sportsmen turned hunts into realms of display and performance. Pedigreed dogs and horses, specialized clothing, finely crafted guns, and paid guides and assistants made their status plainly apparent. Symbolic mastery over trained animals and success in killing wildlife accentuated hunters' authority. In short, hunts amounted to more than mere exercises in recreation and pleasure; they claimed dominance over the natural world and people whom hunters viewed as social inferiors.[9]

Owners of lowcountry estates hunted a wide variety of game. Most initially came for ducks. Throughout the fall and winter months, migrating waterfowl descended on coastal marshes and old rice fields in huge numbers. By the early twentieth century, sportsmen recognized the lowcountry as offering the best duck-hunting on the East Coast. Estate owners also hunted other species, including deer, quail, wild turkey, dove, marsh hens, fox, and opossum. A few owners concentrated on particular varieties of game and others chose not to hunt some because of personal beliefs, but most hunted the full range of wildlife available.[10]

In April 1931, Willis E. Fertig, the owner of Ponemah Plantation on the Black River, touted the quality and diversity of game available. As one of a growing number of estate owners who spent months at a time in the region, Fertig knew the full range of hunting that owners and their guests enjoyed. He named quail, wild turkey, fox, ducks, deer, and raccoon as favored species. "Quail shooting," he observed, "is done from horseback with fast, wide ranging dogs. When the dogs find a covey, the gunners dismount to shoot on the covey rise." Turkey hunts provided special thrills. Fertig viewed them as "perhaps the most fascinating of all the sporting events obtainable." "The wild turkey," he noted, "is a wise and willy bird with a keen sense of hearing and wonderful eyesight, and amply able to take care of himself under all conditions." Fertig considered "the bagging of a turkey . . . a real event." Few forms of sport, he opined, compared with "bringing down a large gobbler." Fox hunts took place at night. Packs of hounds followed the fox by scent "until the fox is either killed or lost." Friends and neighbors typically participated in fox hunts, making them eminently social affairs. Hunters generally brought their own "hounds to add to the pack." Other species pursued nocturnally included raccoon and opossum, which hunters stalked with trained dogs. Deer hunting took place under similar circumstances as fox hunts. Hunters used hounds to "drive the game out of the swamps and thickets [and] past the hunters who are placed on stands to shoot the fleet little animals as they run by at full speed," Fertig explained. "A deer drive," he added, "is a sort of social affair in which friends and neighbors are invited to participate." Afterward, all present enjoyed refreshments.[11]

Fertig's comments not only highlight the conditions that characterized lowcountry gamelands; they hint at the range of associated social possibilities. Different species supplied more than variety; they allowed outings to be tailored to participants' abilities. Quail hunts, as the most challenging of the lot, reserved themselves for skilled hunters who relished the challenge of shooting small, fast-moving birds in flight. Hunting doves and marsh hens offered similar but somewhat less demanding experiences. Duck hunting meant rising before dawn and rowing out to duck blinds on the edge of old rice fields and coastal marshes. Although the sheer number of ducks in the lowcountry generally made it possible for most hunters to obtain a respectable number of kills, conditions did not favor the faint of heart. Duck hunts demanded fortitude and determination—qualities prized by serious hunters. Deer drives most readily accommodated the participation of hunters with modest skills. Even those who never got a good shot could still enjoy the anticipation and excitement that unfolded as deer bounded toward waiting hunters. The camaraderie and socializing that took place afterward provided added benefits.[12]

Symbolically, sportsmen's and sportswomen's practices involved more than met the eye. Informed observers recognized driving as a style of hunting rooted in the sporting traditions of the English gentry. American hunters had also practiced driving, but declining wildlife populations had prompted states in the North and Middle West to outlaw the practice. Driving remained legal in the South, which led hunters to view the region as a bastion of aristocratic traditions. Hunting in the lowcountry thus allowed sportsmen and sportswomen to practice a style of hunting that linked them to British nobles and centuries-old traditions.[13] Even more important, social conditions and the service of African Americans imparted a sense of aristocratic privilege to their activities. Scott E. Giltner has recently shown that white hunters viewed black laborers as essential for hunting in the South. African Americans served sportsmen and sportswomen by performing a host of skilled and menial tasks. African Americans worked as guides; tracked, located, and drove quarry; assisted sportsmen and sportswomen during hunts; readied equipment before outings; and cared for horses and dogs. In some cases, sportsmen and sportswomen admired African Americans' skills as guides. Intimate knowledge of swamps and woodlands made black guides especially effective. Most workers, however, carried out simple tasks that required little skill or expertise. Their labor made it possible for sportsmen and sportswomen to hunt in the manner they expected, without contending with bothersome chores. Regardless of the tasks performed, sportsmen and sportswomen saw black labor as having more than practical value; they saw it as replicating the fabled hierarchies of the antebellum era. As Giltner has written, hunting in the South played to fantasies of "aristocratic mastery." Throngs of black workers affirmed hunters' sense of racial superiority. By echoing the ideal of the "faithful slave" and encouraging hunters to imagine themselves as members of a landed elite, black labor led sportsmen and sportswomen to associate hunting in the South with "racial domination." They saw African American labor as a basic to the experience of hunting in the southern states, just like plentiful wildlife and warm sunshine.[14]

Owners of lowcountry estates had no difficulty obtaining African American workers. As Bernard M. Baruch wrote: "a certain number of Negroes came with the place. They had been born there, as were their fathers before them. They knew no other home." In Baruch's case, about 100 African Americans lived on his lands. They maintained a semi-separate existence, living in three different villages, farming small plots, and hunting and fishing for their own consumption. Many also worked for Baruch on his Hobcaw Barony estate.[15] Similar conditions prevailed on other plantations. Although exact numbers are difficult to obtain, owners' writings and plantation records make clear their reliance on black laborers. At Medway Plantation in Berkeley County, Gertrude and Sydney Legendre had perhaps as many as twenty families living on their lands and a core group of six to eight men who worked as laborers. Photographs of their hunts show African Americans serving as drivers, tending to dogs, and carrying freshly killed game.[16] Farther up the Cooper River at Rice Hope Plantation, Joseph S. Frelinghuysen had "five negro families" living on his lands and employed one "negro houseman" and two "negro maids."[17]

Demographic conditions added to sportsmen's and sportswomen's sense of racial superiority. Throughout the lowcountry, blacks outnumbered whites by large margins. Massive importation of African slaves during the eighteenth and early nineteenth centuries had produced unusually high black-white

Figure 1.1 Hunting party at Mulberry Plantation, circa 1917. Northerners viewed African Americans as fundamental to hunting in the South. Black guides and assistants routinely accompanied sportsmen on hunts. Scenes such as this one abounded at northern-owned sporting estates. Here, hunters talk after a successful hunt while an African American guide holds freshly killed ducks for the camera. *Source:* By permission of Chapman Photo Collection, South Carolina Historical Society, Charleston, S.C.

ratios. On the eve of the Civil War, slaves formed as much as 90 percent of the population in plantation-heavy districts such as Georgetown and Beaufort. After Appomattox, the balance slowly began to shift. As late as 1930, most of the rural lowcountry remained 70 percent or more black.[18] Visitors saw exoticism and "timelessness" as products of longstanding isolation. "Cut off from the outside world by rivers, wide swamps and lack of roads," the lowcountry remained "undisturbed by the restless present," wrote one commentator. Conditions seemed more reminiscent of tropical regions abroad than other parts of the United States. The racial composition of the population appeared especially characteristic of colonial settings. Lowcountry blacks generally had dark complexions and features that whites considered "African." Most also spoke Gullah, a creole language rooted in the unsettled social conditions of the slave trade and early settlement period. Together, these qualities led whites to view lowcountry blacks as a distinct race.[19]

The decaying remains of a massive slaveholding complex further enhanced the sense of exoticism and intrigue. Throughout the lowcountry, deteriorating houses, flooded and overgrown fields, and collapsing barns and sheds recounted a staggering decline. Material decay marked the long-term consequences of the South's failed bid for nationhood and planters' inability to adapt their operations to free labor. Weathered edifices imparted an abiding sense of otherness while simultaneously supplying apparent evidence of the renowned grandeur of the antebellum era. Amid the presence of a seemingly majestic past, sportsmen and sportswomen invariably saw themselves as following in the footsteps of a deposed aristocracy. Ruined plantations provided constant reminders of the region's history, replete with visions of lost gentility, refinement, and authority. For people predisposed toward viewing themselves as an elite, the lowcountry landscape anchored an expansive view of the past that inspired, compelled reflection, and suggested the instability of social and economic power.[20]

In sum, hunting in the lowcountry involved more than mild weather and favorable shooting. Sportsmen and sportswomen found conditions that augmented the status-connoting aspects of recreational hunting and provided uncommonly satisfying experiences. Although northerners had long viewed the South as an exotic land apart, the lowcountry seemed a step beyond. Growing familiarity with the region infused sportsmen's and sportswomen's activities with new meaning. In turn, northerners' winter sojourns assumed new significance. More than simply journeys to enjoy good sport, they became synonymous with vital modes of endeavor. For established elites seeking encounters beyond the norm, the lowcountry offered unique possibilities.

By the time the phrase "plantation life" entered the lowcountry vernacular, pastimes such as horseback riding, boating, sightseeing, golf, and tennis played an important role in estate owners' activities. These pursuits revealed greater social cohesion and stays of longer duration. As the number of people spending the better part of the winter season in the region grew, new social possibilities developed. As longer stays became common, estate owners diversified their activities for the sake of variety, to accommodate bouts of poor shooting, and in response to the varying interests of family and guests. In combination, these trends produced a more elaborate realm of activity. Greater recreational diversity and increased social intercourse merged upper-class social conventions with the social and environmental conditions of the rural lowcountry. The results included greater emphasis on status-connoting behaviors and policing of social boundaries.

Boating, hunting, fishing, sightseeing, and socializing made for demanding routines. As Richard S. Emmet recalled, "life at Cheeha Combahee was strenuous." Days typically started early with fox, duck, or turkey hunts. Owners and guests rose before dawn to get into position by first light. Shoots lasted anywhere from one to three hours. Upon returning from the fields, owners and guests usually ate breakfast before turning their attention to supervising workers or a second round of recreation. At Medway Plantation on the Back River, Sydney and Gertrude Legendre often toured their property on horseback. Other estate owners also made horseback or carriage rides part of their routines. At Mepkin Plantation on the Cooper River, Claire Boothe Luce took guests on carriage rides around the perimeter of the 7,200 acres that she and her husband, the publishing tycoon Henry R. Luce, owned. The trip usually took about four hours.[21]

Lunch provided opportunities for socializing. As Gertrude Legendre noted, she and her husband Sydney often ate midday meals with owners of neighboring estates—"the Chapmans or the Kittredges or the Sharpes." Conversation invariably centered on "plantation life." Sharing meals offered opportunities to discuss the challenges of plantation management, to plan outings, and to exchange notes about shooting conditions. Afternoon activities included fishing and hunting. Members of the Pratt family routinely hunted quail. The Legendres often fished on the Back River and sometimes shot dove with guests. The onset of evening did not necessarily lead to decreased activity. Some owners hunted for raccoon and possum at night. Others spent evenings playing charades, games, or dancing. Owners and guests made use of every waking hour. By filling their days with activity, they made plantation life a vigorous, continuously changing pageant of recreation and entertainment.[22]

Practical needs overlapped with leisure. Boating provides a convenient example. Across the lowcountry, owners used waterborne craft for transportation and pleasure. Jessie Metcalf, owner of Hasty Point Plantation, sped along the Peedee River in a "luxurious speed boat capable of making forty five miles an hour." He and his wife used it to reach hunting grounds, to visit friends and neighbors, and for occasional trips to Georgetown. George Bonbright kept several boats at his Pimlico Plantation "for his pleasure and that of his guests." He found the Cooper River the best way to reach Charleston. In May 1937, the Legendres and Thomas and Alexandra Stone of Boone Hall Plantation took a day-long pleasure trip up the Cooper on the Stone's boat, the "Wampancheone." The group enjoyed stunning scenery, basked in the spring sunshine, and visited Benjamin and Elizabeth Kittredge at Dean Hall Plantation.[23]

As Charleston's popularity as a tourist destination grew, new opportunities followed. The creation of Yeaman's Hall, an exclusive club located twelve miles north of Charleston, brought an influx of visitors from the North and Midwest and expanded owners' social and recreational options. Yeaman's Hall resulted from the efforts of a group of investors led by Henry K. Getchius of New York. The group hired Donald Ross, the leading golf course designer of the era, to create a golf course and commissioned architect James Gamble Rogers to design a clubhouse. The group also built seven guest cottages that collectively offered thirty-three private rooms. Club members also had the option of building private "cottages" on sites scattered across the 1,100-acre property. By November 1931, thirty-three had been built. Yeaman's Hall thus offered options for wintering in the lowcountry and enjoying sport without the obligations of owning a "plantation" and an alternative to the overwhelmingly masculine atmosphere of hunting clubs.[24]

Yeaman's Hall quickly became popular with owners of nearby estates. Joseph S. Frelinghuysen, owner of Rice Hope Plantation on the Cooper River, visited frequently to play golf, dine, and meet friends. The club lay about ten miles south of his estate by river, which made it easily accessible by boat. His neighbor, Henry Luce, built a cottage at the club that he and his family used and which he also made available to members of his staff in New York. The opportunity to "spend three or four days or a couple of weeks there," as Luce wrote to one associate, became a fringe benefit of working for Time-Life.[25] Meanwhile, other estate owners took advantage of similar facilities elsewhere. Sydney Legendre, for example, played golf from time to time at the Summerville Golf Club, a course located about twelve miles west of his plantation.[26]

As plantation life grew to encompass a wider range of recreational pursuits, it also became more exclusive. The proliferation of large estates encouraged development of ties among estate owners, some rooted in existing networks, others entirely new. The Legendres hunted and socialized with the Kittredges not only because their estates lay in close proximity to one another but because of familial ties. Gertrude's parents had known the Kittredges for years and saw them frequently in New York and Palm Beach. When the Legendres' purchased Medway, they reinforced and extended an established relationship.[27] Family connections linked several estate owners. Bayard Dominick, a New York stockbroker, owned Gregorie Neck Plantation on the Coosawhatchie and Tulifinney rivers; his brother, Gayer Dominick, owned an estate on Bulls Island. Financier Franklyn K. Hutton wintered at Prospect Hill Plantation on the Edisto River; his brother, E. F. Hutton, owned Laurel Spring Plantation in Colleton County. Percy Hutton owned Clay Hall Plantation in Beaufort County; his brother, Hans K. Hudson, owned Delta Plantation on the Savannah River. Friendships also formed a basis for plantation ownership. Two friends from Auburn, New York, Charles W. Tuttle and Emerson W. Hitchcock, purchased neighboring plantations on the Black River in Georgetown County, and Cornelius J. Rathborne bought Beneventum Plantation after learning about quality of hunting in the Georgetown area from friend and Yale classmate James P. Mills, whose parents owned Windsor Plantation.[28] Meanwhile, owners of estates along the Cooper River cooperated on projects ranging from construction of a new telephone line to lobbying against a proposed hydroelectric project.[29] In short, existing social connections, shared interests, and ample opportunities to recreate with people from similar backgrounds fostered solidity. As the popularity of the lowcountry grew, estate owners became a distinct community.

Social connections among estate owners accentuated the wealth and power of individual members while highlighting group status. In a penetrating analysis of shooting weekends at country houses in Edwardian England, Mark Rothery has shown how ritualized activities demarcated elite status. Shoots displayed the propertied wealth and status of aristocratic landowners. Fine clothing and firearms, pedigreed dogs and horses, elegant wagons, and a host of other accouterments showed owners' economic resources, access to specialized expertise, and refined tastes. Shoots also served to underpin owners' territorial power by associating their authority with a specific place. Buildings and landholdings invariably demonstrated wealth on an impressive scale. The participation of visitors of high status also impressed locals. The arrival of wealthy, powerful people at an estate caused onlookers to take notice and recognize their lesser standing. Moreover, the social intercourse involved in the hunt delineated social boundaries. Inclusion showed belonging to an elite group while simultaneously identifying others as outsiders. In this manner, shooting weekends defined elite status and the privileges that went with it.[30]

Similar practices undergirded the networks associated with lowcountry estates. For estate owners, hunts offered opportunities to display status and authority among peers. Greater numbers played a crucial role. In the early 1910s, only a handful of northern-owned estates existed in three distinct parts of the lowcountry: near Georgetown, along the Cooper River northwest of Charleston, and in Beaufort and Jasper counties. By the early 1930s, the number had grown dramatically. In the mid-1920s, a veritable estate-making boom began. Northerners suddenly began buying large tracts of land and developing handsome estates. In 1925, northerners bought four tracts centered on old plantations where estate development began immediately. Two more followed in 1926, five in 1927, and another in 1928. By the end of the decade, new estates occupied twenty-six sites scattered across the region.[31]

The estate-making boom resulted in greatly expanded social opportunities. In the mid-1920s, three northern-owned estates lay on the Cooper River between Goose Creek and Moncks Corner. Within a decade, the number grew to thirteen. Similar concentrations developed near Georgetown, on the Edisto River, and in Beaufort, Jasper, and Hampton counties. Overlapping social networks magnified the influence of new estates. Owners introduced visiting friends and family to neighbors, and hunts, horseback rides, and tours of quiet byways supplied ample opportunities for socializing. Participants quickly discovered common interests and shared connections. Meanwhile, the growth of seasonal travel along the eastern seaboard, spurred on mainly by the popularity of Florida resorts, made the lowcountry a convenient stopover point.

Figure 1.2 At the peak of northerners' activities in the 1930s, gatherings reminiscent of summer parties at Long Island country estates became common. This remarkable photo shows a cocktail party in progress on the lawn of Gippy Plantation in 1936. *Source:* By permission of Gaud Family Photo Collection, South Carolina Historical Society, Charleston, S.C.

When one estate owner bemoaned the "open season on plantation owners" that occurred every spring, he highlighted the droves of guests who had to be hosted and entertained. "Anyone that has a car and is tired of the big cities, or does not want to motor north from Florida without breaking [up] the trip" invariably came to stay for a few days.[32]

The growth of northerners' activities brought attention to the lowcountry that highlighted the exclusivity of the new estates and the status of owners and their guests. Lowcountry people learned about owners' activities mainly through word of mouth and from articles in newspapers such as the

Charleston *News and Courier*, the Beaufort *Gazette*, and the Georgetown *Times*. Newspapers routinely heralded the beginning of "the season," when owners returned in the days leading up to the traditional opening of duck-hunting season, and the comings and goings of wealthy visitors.[33] By the late 1920s, the activities of "winter colonists" became well known. Lowcountry people saw estate owners and their guests as part an exclusive world that lay close at hand and yet removed from lowcountry society. Northerners had limited interaction with lowcountry people. Only a few members of old-line families participated in owners' activities, and relatively small numbers of people saw northerners' estates while working as contractors, skilled labor-ers, and deliverymen. In sum, northerners occupied a realm that onlookers knew of, occasionally saw at a distance, and heard and read about, but had no way of knowing for themselves. Proliferating discourses about the new estates made their social and cultural significance plainly apparent.

Northerners' activities further emphasized estate owners' status. Hosting guests invariably placed owners in roles with aristocratic connotations. Whether visitors consisted of close friends and family, owners of neighboring estates, or mere acquaintances, hosting invariably cast owners as masters of a landed estate. Guests invariably recognized owners' authority over domestics and laborers and saw landholdings and buildings as manifestations of wealth. Owners' prerogatives also determined who participated in hunts and other recreational activities. In multiple ways, owners maintained a guise of author-ity rooted in ownership of large landholdings and the deferential behavior of subordinates.

Two organizations underscored the development of the lowcountry sport-ing scene. In March 1932, the founding of the Carolina Plantation Society (CPS) showed new cohesion among owners, the desire of lowcountry people to develop amicable relations with the newcomers, and growing concern about land and wildlife conservation. The CPS resulted from the efforts of Augustine T. Smythe, a Charleston attorney who recognized the growing influence of the "winter colonists" and took up the task of promoting friendly relations between them and lowcountry people. Formed with "nearly forty Northern plantation owners" and a small number of local people, the CPS devoted itself to the "preservation of game in the coastal Carolina area." Annual meetings provided opportunities for members to discuss game conser-vation and land management. The first took place at the Wedge Plantation on March 24, 1934. At least thirteen plantation owners and nine Charlestonians ate dinner and listened to speeches by Archibald Rutledge, the poet laureate of South Carolina, and Charles Lawrence, the owner of White Hall Plantation. Rutledge reputedly expressed "appreciation to various northerners who have restored many historic southern landmarks to their original beauty," while Lawrence discussed "the history of pointers and setters as hunting dogs."[34]

Subsequent meetings followed a similar pattern. In 1936, 1937, and again in 1938, members and guests met at a northern-owned estate to eat, socialize, and talk. On January 5, 1936, forty-five people gathered at Medway Plantation to elect officers and hear Zan Heyward, chairman of the South Carolina Game and Fish Commission, discuss game preservation. Two months later the group lunched and listened to four speakers at William R. Coe's Cherokee Plantation. In 1937 the CPS met at Friendfield Plantation on the Sampit River. Year after year, members and small numbers of guests came together to socialize, exchange notes about plantation management, and learn about game preservation.[35]

Although Smythe intended the CPS to promote interaction between plantation owners and lowcountry people, it proved most effective at forging solidarity among estate owners. Owners consistently made up the largest group of participants. At the 1934 meeting, for example, thirteen owners, nine Charlestonians, and three guests attended. The following year, the organization's membership consisted of twenty owners and eight Charlestonians. In 1937, thirteen owners, eleven Charlestonians, and twelve guests attended the annual meeting.[36] Owners benefited from opportunities to socialize, to discuss common concerns, and to learn about game conservation. Since CPS meetings brought together owners from different parts of the lowcountry, they encouraged interaction among people who might otherwise have had little or no contact. In this sense, the CPS helped plantation owners see themselves as a cohesive social group. By delineating shared interests and promoting a region-wide identity, the CPS fostered unity beyond existing norms.

The CPS did less for lowcountry people. Despite Smythe's intentions, the group had relatively little influence. Only a small cohort of Charlestonians joined and attended meetings. They included Smythe; Charles L. Mullally, a real estate broker who catered to prospective plantation buyers; E. Milby Burton, the director of the Charleston Museum; and several members of patrician families. Other lowcountry people participated intermittently, but only in small numbers. Thus, the number of lowcountry people who interacted with estate owners through the CPS never exceeded more than a handful. Rather than facilitating interaction with large numbers of people, the CPS served a small number of elite actors, some of whom already had access to winter colonists. In this sense, it did relatively little to promote friendly relations between the two groups.[37]

The second organization that showed the growing strength and cohesion of estate owners combined training of hunting dogs with socializing and recreation. In the spring of 1932, James Kidder, the owner of Green Pond Plantation in Beaufort County, struck upon a plan to bring about "a higher degree of understanding" and a "closer social association between the plantation owners and their families." On November 10, 1932, thirteen estate owners

Figure 1.3 The Carolina Plantation Society demonstrated growing solidarity among estate owners and recognition of common interests. This photograph shows the organization's 1938 meeting in progress at Harrietta Plantation near Georgetown. The speaker may be Howard S. Hadden, owner of Springbank Plantation in Williamsburg County, who according to the Charleston *News and Courier* briefly addressed the group about the South Carolina Federation of Agriculture, Commerce, and Industry. *Source:* Courtesy Charleston Museum, Charleston, S.C.

met to establish the Plantation Owners' Bird Dog Association (POBDA). The organization made development of better bird dogs its primary goal. From the beginning, however, it also had a strong social orientation. Members saw it as a vehicle for organizing recreational activities, sharing information, and promoting neighborly relations. Like the CPS, the POBDA became central to estate owners' awareness of common interests. In the case of POBDA, its reach proved more localized, extending only to owners in the southeastern corner of the state, in Beaufort, Hampton, Colleton, and Jasper counties. Still, for "winter colonists" in these areas, it showed their growing numbers and established events that became a highlight of the winter season.[38]

Field trials became POBDA's major activity. Beginning in February 1933, members gathered annually at a northern-owned estate or hunting club to watch dogs compete in a series of tests. Judges evaluated dogs on physical attributes, appearance, training, their ability to follow commands, and their performance at common tasks such as retrieving game and flushing coveys.[39] Field trials are part of a longstanding tradition of elite hunting. Usually hosted by kennel or hunting clubs, trials assess training, breeding, grooming, appearance, experience, and obedience. Trials are more demanding than actual hunts. Judges evaluate form and style as well as performance. Subservience to handlers and proper handling responses are also prized. Field trials thus emphasize breeding, grooming, and the fine points of training more than competency. For these reasons, they appeal to owners and trainers of pedigreed dogs.[40]

POBDA's field trials demonstrated northerners' growing commitment to the lowcountry. According to one account, northerners used poor quality hunting dogs early on. "Time was when the 'grass prowler' reigned supreme," noted *Country Life* magazine. "Native-bred, 'nigger-broke' dogs" got the job done, but just barely. They moved slowly, never exerted themselves, could usually find and retrieve game but "neither thrilled nor inspired." Such "plodding laborers" satisfied hunters' basic needs but nothing more. As sportsmen's and sportswomen's interests in the lowcountry grew, they sought out better dogs. POBDA's field trials allowed owners to evaluate their dogs' abilities and compare them to those of friends and neighbors. They also presented well-groomed-and-bred dogs to peers. As yet another opportunity for displaying specialized knowledge and evidence of wealth, field trials underscored the exclusivity and status display of elite hunting.[41]

The POBDA field trials quickly became a celebrated affair. Running the trials took "the greater part of two days." In addition to the various tests for the dogs, they included plenty of feasting and socializing. Attendees turned out in their "best bib-and-tucker," eager to enjoy "the camaraderie of the shooting field." One day featured a barbecue prepared by "Gullah cooks." As an "annual 'get-together' among the plantation owners, their families, guests and employees," the trials fostered "a spirit of genuine good-fellowship."[42]

In different ways, the CPS and the POBDA demonstrated the strength of the social networks associated with northern-owned estates and the growth of owners' interests. Both organizations showed close bonds among estate owners, owners' sense of belonging to a cohesive group, and deep interests in the lowcountry. Both reflected the proliferation of northerners' estates and the social and recreational opportunities that resulted. Moreover, both organizations showed progress beyond the informality of earlier years. In the space of barely two decades, large estates had become the focus of the lowcountry sporting scene. Coordinated activity demonstrated the role that seasonal visits to the lowcountry played in owners' lives. As owners' interests in the region deepened, organizing, socializing, and recreation took on new dimensions. Coupled with ad-hoc forms of mutual assistance, the CPS and the POBDA showed how far northerners' activities had moved beyond the comparative simplicity of earlier years.

If vigorous activity and elite social networks supplied plantation life with essential attributes, the social and environmental conditions of the rural lowcountry rounded out its character. Large numbers of "plantation negroes," the semitropical environment, and decaying remains of earlier periods imbued northerners' experiences with exoticism and intrigue. Interactions with lowcountry blacks proved especially important, for they affirmed owners' and guests' sense of racial superiority. Routine exchanges and ritualized activities supplied northerners with what they saw as evidence of African Americans' moral, intellectual, and physical inferiority. Plantation life continually reified northerners' racism while also accentuating cultural differences. By reminding owners and guests of their outsider status, it accentuated the role of the lowcountry in their lives and sharpened individual and collective identities.

Owners and their guests saw African Americans as basic to plantation settings. Just as black labor seemed essential to hunting in the southern states, northerners saw African Americans as a necessary part of any plantation environment. The combination of black bodies, labor, and white authority recalled the supposedly harmonious race relations of the antebellum era and mirrored conditions across the contemporary South. Sydney Legendre relished managing "the men." According to his wife, he "loved bossing everyone around."[43] One visitor to Boone Hall Plantation commented on the number of laborers she observed and their busyness. "One great charm about Boone Hall is its great activity," she wrote. "There seems to be so much doing every place you go. About a hundred negroes are employed just now, so you are constantly seeing someone raking, cleaning, picking nuts, planting, or doing something else."[44] Although Boone Hall, which the Stones ran as commercial farm, employed more workers than most northern-owned estates, the basic point remained valid. For many whites, white supervision of black labor seemed characteristic of plantations. Claire Boothe Luce made the point even more clearly in an article published in *Vogue* magazine in 1937. Speaking of her Mepkin Plantation, she asked rhetorically, "You know what it does to you to know you own as far as the eye can see, and to have twenty black plantation hands—or rather black faces—you can't tell one from the other—bow as you pass, and say, 'Yas, suh, Boss?' I'll tell you. It makes you feel like a 'Southern aristocrat.'" She doubted that southerners took such behavior as "a compliment," but she did. A new plantation owner "just *loves* to get that feeling," Luce insisted.[45] For her and other white Americans of the era, black deference seemed as innate to plantations as warm weather, cotton, and cornbread.

Black workers did more than simply affirm owners' sense of racial authority; they caused endless frustration. In this sense they made plantation ownership a challenge—work—and countered its image as a leisured indulgence. Just as antebellum planters found the apparent laxity and indolence of their slaves endlessly aggravating, owners of sporting estates saw their

Figure 1.4 Supervising plantation laborers affirmed owners' authority and highlighted the responsibilities associated with estate ownership. In this photo, Clarence E. Chapman (at left), owner of Mulberry Plantation, looks on as a white foreman and black laborers finish construction of a trunk, a type of floodgate used to regulate water levels on tidal rice fields. *Source:* By permission of Chapman Photo Collection, South Carolina Historical Society, Charleston, S.C.

laborers as slothful and incompetent. Luce did not mince words when she described "plantation domestics" as "lazy." She related a series of incidents that convinced her of the impossibility of running Mepkin efficiently. On one occasion, a laundress had begged to go home because of "trouble in mah feet." In another instance, a stable boy refused to come to work after having been "powerful trubbled all night by a hant." Superstitious behaviors and a general aversion to moving quickly caused Luce constant aggravation. "The normal, accepted, and miraculously efficient tempo in which things are done 'up North'—just simply *does not* operate in the South," she explained. "Southerners," she declared, "are a race of people who do things slowly."[46] Sidney Legendre voiced similar complaints about his workers. "Negroes are unable to do things for any length of time," he declared after a particularly frustrating episode. "They work only when there [*sic*] interests are aroused, and this state is not an enduring one," he lamented. On another occasion, during an especially cold spell, workers peppered him with what he considered excuses for not working. "Boss it's so cold I can't hold this shovel no how," he recounted one saying.[47]

Owners also encountered difficulty with plantation superintendents. Most owners hired a white male to manage their estates. When everything went smoothly, owners rarely mentioned these men. When owners found cause for dissatisfaction, they complained vehemently. In February 1940, Legendre questioned the ability of his superintendent, Waring Bunch. Legendre considered him "willing" but inept. He "does not know how to do things," Legendre complained. Bunch did "not realize that things must be neat, that machinery should be under the shed, paper picked up and the place kept spotlessly." Pondering the situation, Legendre asked rhetorically, "Will a new man? That is the question."[48]

For months, Legendre considered his options. "The question of an overseer is one that takes up most of my waking thoughts," he noted in February 1940. "Good men are difficult to find," he lamented. Speaking of a problem he saw as endemic to owners of large estates, he observed, "One may always secure a broken down southern gentleman who wishes to while away the remainder of his life at your expense shooting and hunting on your place under the guise of being an overseer." Getting "a more educated man with vision and taste," however, was a different matter. In the end, Legendre interviewed several men but never found one to his liking. He and Gertrude struggled with laborers and a supervisor whom they viewed as inadequate. The results included a litany of complaints and, according to Sidney, "a certain messy atmosphere" and "feeling of unfinishment" that he saw as indicating how much remained to be done.[49]

Owners who hired northern men as overseers did not necessarily fare any better. While some owners considered native southerners best suited for overseeing plantation operations, others disagreed. Joseph S. Frelinghuysen, for example, employed a northerner, Herman Hansen, as his superintendent. In the spring of 1931, Frelinghuysen became concerned about Hansen's management of the property, particularly his relations with subordinates and others parties. In a frank and forthright letter, Frelinghuysen declared that he would no longer permit "constant turmoil and dissention." He recalled a litany of problems that had arisen. "You and McGinnis quarreled; you quarreled with the Captains on the boat; with Myrtle; with the men who took care of the horses; with Pennington; Mrs. Murray; with Katherine and Lowell; and now with Mercer." "I am just tired of it," Frelinghuysen explained. He insisted on having "absolute agreement as to the management of the plantation" or Hansen's resignation. "The matter is entirely up to you," he added. Hansen apparently complied with his employer's wishes, for he stayed on as superintendent for another several years.[50]

The frustrations that Legendre and Frelinghuysen experienced demonstrate the seriousness with which owners viewed the management and development of their estates and expected to see measurable progress. Historians have tended to view country estates principally as indulgences. Their basic purpose, form, and associated activities embodied the social and cultural priorities of a class with unprecedented wealth and the desire to display it conspicuously. Hence the emphasis on sumptuousness, grandeur, and aesthetic pretention. As stage sets for the antics of social and economic elites, country estates displayed owners' wealth and authority with aplomb. Yet for many owners, the workings of their estates proved as important as outward appearances. The two went hand in hand; separating one from the other is virtually impossible. Yet neither is it possible to overlook the effort and energy some owners put into improving, maintaining, and ensuring the smooth operation of their estates. The potential for these activities to yield satisfaction and a sense of accomplishment captivated the imaginations of some owners, and when all did not go as planned, disappointment resulted. As the examples of Legendre and Frelinghuysen demonstrate, owners took their self-assigned responsibilities seriously and sought to rectify problems when they arose. Presiding over their estates contributed to their sense of self and projected authority of a kind derived from supervising subordinates and directing the workings of a complex enterprise.

Other forms of interaction with laborers yielded markedly different results. Northerners' fascination with lowcountry blacks' seemingly primitive behaviors placed black workers and estate owners and their guests in new roles. Although basic relationships between the two groups owed a great deal to established patterns, new modes of behavior also developed. Ritualized performances supply a conspicuous example. Throughout the region, African Americans engaged in practices that whites viewed as manifestations

of longstanding traditions. Singing traditional music and holiday gatherings proved the most common. As events that placed African Americans
in prescribed roles and saw northerners enjoy the benefits of estate ownership, such occasions highlighted the unequal power relations that prevailed
on northerners' plantations and the northerners' desire for status-affirming
activities. Like their antebellum predecessors, owners of sporting estates
sought validation of their wealth and social status.

Consider, for example, the ritual that took place at Medway Plantation on
Saturday evenings. Soon after dark, several African Americans would gather
in a yard beside the main dwelling, where the Legendres and any guests
who happened to be staying with them sat by a fire, talking and enjoying the
evening air. The African Americans arranged themselves in a row and began
signing. One would lead: "Who buil' duh Aa'k?" In deep, rich voices, the
others would respond: "Norah, Norah Lawd." As the song progressed, the
group began to sway and move, clap their hands, and harmonize. The soaring
depth and power of their voices offered an unencumbered view of the beauty
and texture of African Americans' folk traditions. Increases in tempo produced commensurate quickening of vocal expression and bodily movements.
Some songs culminated in dramatic crescendos of physical and vocal expression. No doubt the Legendres and their guests clapped, smiled, hummed, and
perhaps even sang along while taking in the performance. According to one
report, the practice of singing spirituals for plantation owners represented
"an old Low Country tradition." Owners undoubtedly felt privileged to
experience it firsthand. For them and their guests, it represented a powerful
expression of African American culture and a direct legacy of slavery and the
isolated conditions that had prevailed in the rural lowcountry since.[51]

By the early 1930s, spirituals had become a celebrated tradition. Observers near and far viewed them as a vestige of a lost era, a living link to the
plantation culture of the Old South. In the lowcountry, the Society for the
Preservation of Negro Spirituals (SPNS), an organization made up mostly of
descendants of slaveowners, promoted the music by transcribing, recording,
and performing it "with affectionate fidelity." SPNS members performed
spirituals while wearing clothes typical of antebellum slaveowners. Thus,
they assumed the persona of masters, not slaves or contemporary African
Americans, a move that posited white elites as authorities on African
American history and culture. The SPNS achieved impressive popularity.
As Charleston became a leading tourist destination, the group performed
extensively in the city, elsewhere in the lowcountry, and toured outside of
the South in 1929, 1930, and 1935. The latter tour included a concert at the
White House, where Franklin and Eleanor Roosevelt, members of Congress
and the cabinet, and other distinguished guests heard the group sing and
explain the role of spirituals in the heritage of the lowcountry.[52]

Estate owners viewed spirituals as one of the lowcountry's most colorful traditions. Although no record of the Legendres attending a SPNS performance exists, it is possible that they did. Other northern plantation owners attended SPNS concerts and supported the organization's activities through memberships and by purchasing copies of *The Carolina Low-Country,* a mammoth collection of historical essays, poems, and transcribed spirituals complied by the SPNS and published by the Macmillian Company of New York.[53] Moreover, spirituals received considerable attention in the northern press. Accounts in leading newspapers and magazines presented spirituals in much the same way as the SPNS did. An article published in *Country Life* in 1935, for example, characterized spirituals as an "exotic music" that survived from the era of slavery, "a heritage we must not lose."[54]

Practices of the kind that occurred on Saturday nights at Medway took place throughout the lowcountry. African Americans routinely sang spirituals for northerners at other plantations. In 1935, for example, the *News and Courier* reported on a performance near Dean Hall Plantation. According to the newspaper, "negro plantation hands" sang "in the moonlight" for a crowd gathered around a small bonfire. Attendees noted that "the negroes' voices had a mysterious African aspect."[55] Similar performances occurred elsewhere. At the Oaks Plantation in Goose Creek, "the Negro farm-hands" sometimes gathered outside the main house "on moonlight nights." There, they regaled "the 'company' with 'shouts' and 'singing games' and 'buck and wings.'" According to one visitor, these episodes took place "quite in ante-bellum fashion."[56] Similarly, Richard Emmet recalled evening hunts at Cheeha Combahee Plantation including bonfires on the road and "members of the local population singing spirituals and telling stories."[57]

Other ritualized practices provided similar forms of interaction. As with spirituals performances, many echoed practices common under slavery. At Springbank Plantation in Williamsburg County, retired insurance magnate Howard S. Hadden and his wife staged Christmas celebrations for their laborers and other African Americans from the surrounding area. "Every Christmas Mr. and Mrs. Hadden delight in playing Santa Claus to the negroes," the *News and Courier* reported. It described the 1937 celebration as follows:

> One old 'uncle' is dressed up as Saint Nick. He stands pompously before the tall columns of 'Springbank' and claps his hands as a gift is handed to each member of his race as they file past. The line this Christmas was just 260 darkies long. And not a one left empty handed. From the veranda and the lawn one hundred fifteen white guests looked on while the negroes sang spirituals and danced.[58]

The Haddens' version of plantation life replicated a well-known ritual of the antebellum era. At plantations across the South, slaveowners had observed Christmas by giving gifts to slaves and allowing a day of rest. In casting themselves as Mr. and Mrs. Claus, the Haddens emphasized their power and authority. Gift-giving, in this sense, displayed wealth and status. It emphasized the Haddens' role as landowners and employers, highlighted the apparent deference of hundreds of African Americans, and carried out the entire spectacle before an audience of 115 white guests. Rarely if ever did an estate-owning couple make their authority so clear.[59]

At other estates, holiday celebrations assumed somewhat different contours. In January 1935, the *News and Courier* reported on a New Year's celebration at Bleak Hall Plantation. According to the newspaper, when Dr. James Cowan Greenway and his wife arrived at Bleak Hall soon after Christmas, "the negroes" immediately "appeared in a body at the maser's cottage." Each held "a chicken for the doctor and his wife." Impressed by the thoughtfulness of their "people," the Greenways decided to throw a picnic. The event took place several days later. "Everyone on the plantation" attended. The Greenways served cake and ice cream and gave every woman present five dollars. "One old man," the paper reported, enjoyed himself so much that "he ate seven ice cream cones in spite of the rather chilly weather."[60]

At Boone Hall Plantation, Thomas and Alexandra Stone organized a similar gathering. On May 17, 1937, Thomas Stone noted in his journal, "We are having a party for all the colored people on the plantation—and, I expect when the word gets around, for all the colored people in the Parish." He and Alexandra planned to serve "mulatto rice, beer, and pop." For entertainment, they hired the Jenkins Orphanage Band, Charleston's best-known African American musical troupe. Two days later, Stone reported the party to have been "a great success." About 150 people attended—well over the fifty he had expected. "Everyone seemed to enjoy themselves enormously," he noted. Singing and dancing took place, and "the mulatto rice, cake, ice cream, beer, soft drinks and candy disappeared like the dew on a hot summer morning." A lightwood fire and a full moon supplied ample illumination. According to Stone, the "whole scene . . . was most entrancing." He added that "absolute perfection was missed only because of the rank smell from about 100 2½ cent cigars being smoked furiously at the same time."[61]

Other activities also highlighted owners' authority. Some estate owners showed concern for the well-being of their laborers and tenants by erecting new houses and other buildings for their use. At Arcadia Plantation on the Waccamaw Neck, Isaac Emerson built a church, a school, and an infirmary "for the negroes" living on his lands. At Mepkin Plantation, the Luces built several cabins for their laborers. Brick construction and indoor plumbing and heating made them superior to the vast majority of housing in Berkeley County and far above the norm for African Americans throughout the lowcountry. Bernard Baruch provided health care and education for his employees and their children. He operated two schools at Hobcaw Barony, one for the children of black employees, the other for those of white employees. Other plantation owners may have assisted blacks in other ways. Harry F. Guggenheim, the owner of Cainhoy Plantation north of Charleston, considered making a monetary gift to Engenie Broughton, an African American nurse employed by the South Carolina Department of Health. Broughton apparently requested assistance from Guggenheim's superintendent, V. C. Barringer, who viewed her need as potentially "worthy." No record of Guggenheim's decision survives, but correspondence indicates that Barringer urged his employer to consider it and took steps to determine its merits. Barringer insisted on speaking with Broughton personally, for he worried that outright support would mean "there will never be any end to this with her and the other colored people."[62]

Estate owners' philanthropy extended to white institutions. In at least two instances, owners contributed to the rehabilitation of rural church buildings. Dr. Henry Norris and his wife, owners of Litchfield Plantation, gave a font made of Italian marble and paid for construction of a brick wall and iron gate at All Saints Waccamaw Church on Pawley's Island. Harry Guggenheim undertook a more substantial renovation of the St. Thomas and St. Denis Church near Cainhoy. According to a 1936 report, the building had been "for many years falling into ruin." Guggenheim hired a local contractor to repair the roof, windows, and brickwork. In other instances, estate owners sought to revitalize lagging congregations. In 1925, Joseph S. Frelinghuysen encouraged his friend and neighbor George Bonbright to attend services at Strawberry Chapel. "We are trying to make a congregation so that we can hold services there every Sunday, or at least twice a month," he explained. "The church is a very historic one but is poor," Frelinghuysen noted. "All of the sportsmen on the [Cooper] River are backing me in an effort . . . [to] have the services more frequently."[63]

Support of laborers, residents, and lowcountry whites cast estate owners in the role of patrons of the needy and less fortunate, in a manner seemingly reminiscent of the noblesse oblige commonly associated with antebellum planters. By aiding people on whom they depended or wished to draw into friendly relationships, estate owners cultivated beneficial ties at limited expense. As yet another means of demonstrating wealth and status, philanthropy added to owners' image as an elite with virtually limitless resources and genuine concern for lowcountry people. Patronage did nothing to alter the gross inequalities at the heart of lowcountry society. Rather, by reinforcing patterns of deference and subordination, it upheld Jim Crow and longstanding class divisions. Northerners benefited from maintenance of such hierarchies. By keeping lowcountry people eager for activity and investment in rural areas and large numbers of laborers willing to work for low wages, they supported the creation and operation of large estates.

Plantation life lasted only a short while. During the 1930s, several developments undermined the conditions that northerners prized. New Deal programs dealt the first blow. African Americans flocked to job opportunities with federal relief programs, triggering an exodus from rural areas that compromised northerners' ability to maintain their estates. As workers left sporting plantations in ones and twos, owners and superintendents fretted about their ability to carry out routine maintenance and support hunting and other forms of recreation. Soon, they found themselves struggling to do both. New Deal programs also brought a surge of activity to the countryside that disrupted conditions on northerners' plantations. Construction of new roads, bridges, and public buildings put a "modern" face on many communities and heralded the advent of closer communication with major cities and the world beyond. Rural electrification, soil conservation, and agricultural assistance programs introduced similar changes elsewhere. In all, the New Deal brought a swift and sudden end to the sense of isolation that had long prevailed, destroying the remoteness and serenity that northerners coveted.[64]

Changes in upper-class culture also played a role. The rise of "café society," an international mélange of celebrities and socialites that first attracted attention in the late 1920s, evinced the rise of a more cosmopolitan lifestyle and the waning influence of older forms of display. Café society reputedly developed when some socialites became "bored" with Newport and "stuffy dinners off gold plates." Looking to Times Square and Hollywood for entertainment, they soon found "wittier, more stimulating . . . playmates" in writers, artists, and stars of film and radio. Café society marked the rise of the entertainment industry as a major force in American life and the growing role of wit and banter in mass culture. Above all, café society signaled the decline of rigid exclusivity and conscious aping of European nobles. Although older modes of behavior did not disappear overnight, they nonetheless lost currency. Country estates, once anchors of elite society, declined in influence. Lowcountry sporting plantations, as a variant of the type, did also.[65]

World War II marked a turning point. The changes set in motion by the New Deal accelerated dramatically after the Japanese attack on Pearl Harbor. Military service and employment in defense industries took remaining laborers off plantations, and several estate owners also contributed to the war effort. Civil defense measures and the expansion of the Charleston Naval Yard gave the war a presence throughout the region. The completion of the Santee-Cooper project, a massive hydroelectric complex funded by the New Deal, transformed large swaths of Berkeley and Charleston counties. Moreover, the shift in racial attitudes sparked by the war made many of the rituals common on northern-owned plantations less palatable. Owners recognized the significance of these changes immediately. Surveying their influence in the spring of 1941, one remarked, "This was the end plantation life."[66]

By the late 1940s, only vestiges of the sporting scene that had existed in earlier years survived. Northerners continued traveling to the lowcountry annually to hunt and relax at large estates, and shooting parties, hosting friends and family, and varied forms of recreation remained common. Yet the sense of isolation, remoteness, and exoticism—the mystique of a region mired in a different time, seemingly untouched by modern life—had gone. Remnants remained, but only in scattered form. Moreover, shifting patterns of ownership and generational change eroded estate owners' cohesiveness as a social group. Most owners had reached middle age by the time they created their estates, and some did not live long afterward. Throughout the 1930s and 1940s, obituaries announcing owners' deaths appeared in major newspapers. Although exact figures are elusive, a sample of twelve owners for who detailed biographical information is available reveals that five died in the 1930s and five in the 1940s. Sales of large estates accelerated during World War II and the years that followed. Of a sample of fifteen plantations, twelve old during the 1940s, mostly to non-northern interests. As the *New York Daily News* reported in 1944, "the very latest trend in plantations is that they are reverting back to the natives." Declining enthusiasm for the lowcountry led owners and heirs to sell at diminished prices, often far less than had been invested. Buyers tended to be prominent businessmen from coastal towns and members of old-line families. In 1942, the International Paper Company purchased Rice Hope Plantation from Willis E. Fertig; three years later a lumber company from Conway, South Carolina, purchased Robert Goelet's Wedgefield Plantation. In 1947, H. O. Schoolfield, a businessman from Mullins, purchased Hasty Point Plantation from Jesse Metcalfe. The following year, two Conway businessmen, Charles N. Ingram and E. Evan Dargan, purchased Nightingale Hall, the former estate of J. S. Holiday. As the years passed, wealthy "winter colonists" became less prevalent and estate owners became a more heterogeneous group.[67]

Viewed in perspective, plantation life stands as a short-lived phenomenon that merged traditional field sports with the social Darwinism and racial thought prevalent among upper-class Americans in late nineteenth and early twentieth centuries. As historians such as Michael Kimmel, Gail Bederman, and E. Anthony Rotundo have shown, concerns about the apparent feminization of American society surged in the 1890s and continued in later decades. The loss of male autonomy, the arrival of millions of immigrants, new feminist movements, and concerns about overcivilization led many to fear a loss of manly vigor. In response, men reasserted manhood through sport, strenuous exercise, outdoor pursuits, and a host of other activities. Although historians have tended to emphasize the masculinities of middle- and working-class men, the upper classes also feared loss of manliness and strength. In fact, upper-class men saw themselves as especially susceptible to feminization.

Wealth largely negated concerns about loss of economic independence, but sumptuous living offered as clear an example of overcivilization as any. Lavish balls, opulent parties, around-the-clock servants, rich and plentiful food, comfort taken to excess—all showed the fruits of exceptional wealth. All showed abandonment of the thrift and industry practiced by earlier generations. All suggested indulgence taken to excess.[68]

Generational influences added to the problems faced by upper-class men. Heirs of great wealth struggled to match the accomplishments of their forebears. Who could possibly achieve the success of a Vanderbilt or a Carnegie, let alone assemble a comparable fortune? Comfortable childhoods tended to produce limited ambitions, not drive and determination. Worse, overindulgence often led to instability later in life. As scholars such as Frederic Cople Jaher have observed, sizeable inheritances sowed behaviors that led to profligacy, drunkenness, divorce, neurosis, and no shortage of other problems. In short, heirs struggled to exercise responsibility and self-control. Without the imperatives imposed by economic necessity, virtue proved elusive.[69]

Plantation life represented one response to the anxieties of a democratizing age. As gratuitous display fell into disfavor and concerns about concentrated wealth continued to grow, upper-class Americans redoubled their efforts to cultivate social and cultural authority. Lowcountry estates became one manifestation of these impulses. Although their origins lay in pursuit of wildlife and undeveloped land, their eventual role as settings for ritualized performances gave them a specialized place in the lives of a small elite. The range of activities they supported and the manner in which they cast owners and guests differentiated them from comparatively modest hunting retreats and hunting clubs. Aesthetically, northerners' plantations exhibited the dignity and reserve commonly associated with the behavior of gentlemen. Eschewing ostentatious display, they embraced traditional modes of design, stylistic conservatism, and evidence of time's passage. As social and recreational venues, they afforded rare opportunities for wealthy men and women to demonstrate ability while surrounded by status-affirming symbols. Hunting represented one of the most powerful forms of exploit available, and prowess in the field supplied ample evidence of accomplishment and success. Wild game taken in a sporting fashion, in keeping with all the conventions that elevated hunting from subsistence to sport, posited hunters as masters of nature, knowledge, and self-restraint. For men and women eager to display authority, it proved exceptionally potent.

The lowcountry accommodated sportsmen's and sportswomen's interests with ease. Time-honored traditions, the decaying remains of an enormous slaveholding complex, and the survival of locally distinctive cultures made an authentic past a powerful presence. Subtropical flora and fauna, near-endless expanses of coastal marshes, and the "Africanness" of lowcountry blacks imparted an exoticism akin to distant lands abroad. Hunting of unusual quality and exchanges with cultural "others" supplied a continual stream of vital experiences. For status-conscious men and women seeking affirmation without undue risk or strain, the region proved eminently satisfying. Simultaneously a land apart and yet readily accessible, it indulged lusts for adventure and intrigue, all within a day's journey of the major cities of the East Coast.

In the years between the world wars, wealthy sportsmen and sportswomen turned old plantations into exclusive realms of performance and display. The sporting plantations of the Carolina lowcountry occupied a quasi-colonial milieu seemingly removed from the world at large and steeped in a distinctively American past. Plantation life supplied limitless opportunities for adventure, exotic encounters, camaraderie, and "good sport." As a sphere where traditional pastimes and unchallenged social hierarchies connoted status, plantation life offered people of wealth and privilege opportunities to demonstrate ability among peers virtually without risk. In an era of mounting challenges from below, it reassured, affirmed, and entertained. Moreover, it underscored the dramatic changes that had unfolded with the making of new sporting estates. At the beginning of the twentieth century, plantations throughout the lowcountry had lain ruined and abandoned, their future uncertain and apparent usefulness spent. Now, dozens had become winter residences and private shooting retreats for members of a privileged elite. Plantation life evinced the myriad changes that had occurred. Neither a simple mix of upper-class pastimes nor a conscious bid to live out an imagined plantation past, it blended the exigencies of the moment with conditions specific to the rural lowcountry. The results spoke for themselves.

NOTES

1. Chalmers S. Murray, "Attracted by Climate Northerners become Land Barons of Carolina Coast," *News and Courier* (Charleston, S.C.), Dec. 15, 1929, p. A3; Chlotilde R. Martin, "Low-Country Plantations Stir as Air Presages Coming of New Season," *News and Courier*, Oct. 16, 1932, p. A6; Laura C. Hemingway, "Williamsburg Boasts Many Game Preserves for Hunt," *News and Courier*, Jan. 11, 1937, p. 3; "Wealthy Hunters at Williamsburg," *News and Courier*, Feb. 6, 1938, pp. B3, B10; James C. Derieux, "The Renaissance of the Plantation," *Country Life* 41, no. 3 (Jan. 1932): pp. 34–39; George C. Rogers, Jr., *The History of Georgetown County, South Carolina* (Columbia: University of South Carolina Press, 1970), chap. 21; John H. Tibbetts, "The Bird Chase," *Coastal Heritage* 15, no. 4 (spring 2001): pp. 3–13; Daniel J. Vivian, "The Leisure Plantations of the South Carolina Lowcountry, 1910–1940" (Ph.D. diss., Johns Hopkins University, 2011).

2. Martin, "Low-Country Plantations Stir"; Derieux, "The Renaissance of the Plantation"; Rogers, *History of Georgetown County*, chap. 21; Vivian, "Leisure Plantations of the South Carolina Lowcountry," chaps. 1–4.

3. On the "second Yankee invasion," see Rogers, *History of Georgetown County*, chap. 21; Tibbetts, "The Bird Chase"; Vivian, "Leisure Plantations of the South Carolina Lowcountry," chaps. 1–2. On the rise of tourism in the lowcountry, see Walter B. Edgar, *South Carolina: A History* (Columbia: University of South Carolina Press, 1998), p. 493; Walter J. Fraser, Jr., *Charleston! Charleston!: The History of a Southern City* (Columbia: University of South Carolina Press, 1989), pp. 368–375, 380–381, 386; Stephanie E. Yuhl, *A Golden Haze of Memory: The Making of Historic Charleston* (Chapel Hill: University of North Carolina Press, 2005); Michael N. Danielson and Patricia R. F. Danielson, *Profits and Politics in Paradise: The Development of Hilton Head Island* (Columbia: University of South Carolina Press, 1995); Barbara F. Stokes, *Myrtle Beach: A History, 1900–1980* (Columbia: University of South Carolina Press, 2007).

4. Northerners' memoirs frequently mention plantation life. See Gertrude Sanford Legendre, *The Time of My Life* (Charleston: Wyrick and Co., 1987), p. 70; Francis Cheston Train, *A Carolina Plantation Remembered: In Those Days* (Charleston: History Press, 2008), chap. 4. For contemporary mentions, see Benjamin R. Kittredge to S. Dana Kittredge, n.d. [ca. 1927], fol. 4, box 15, Kittredge Family Papers, South Carolina Historical Society, Charleston, S.C. (hereafter SCHS); Martin, "Low-Country Plantations Stir"; "Wedgefield Plantation," *Country Life* 75, no. 2 (Dec. 1938): p. 53. An earlier account that describes activities characteristic of plantation life but does not use the term explicitly is Murray, "Attracted By Climate."

5. On elites yearnings for vital experience, see T. J. Jackson Lears, *No Place of Grace: Antimodernism and the Transformation of American Culture, 1880–1920* (New York: Partheon Books, 1981).

6. Vivian, "The Leisure Plantations of the South Carolina Lowcountry," pp. 16–35.

7. On the decline in waterfowl populations, see Henry H. Carter, *Early History of the Santee Club* (Boston[?]: n.p., 1934), pp. 11–15; Suzanne Cameron Linder, *Historical Atlas of the Rice Plantations of the ACE River Basin – 1860* (Columbia: South Carolina Archives and History Foundation, Ducks Unlimited, and the Nature Conservancy, 1995), pp. 309–310; "Duck Season Finished," *News and Courier*, Feb. 1, 1930, p. 2; "Big Duck Hunting Season Expected," *News and Courier*, Nov. 14, 1933, p. 2; "Duck Season Ends Tomorrow," *News and Courier*, Dec. 18, 1935, p. 14.

8. Daniel Justin Herman, *Hunting and the American Imagination* (Washington, D.C.: Smithsonian Institution Press, 2001), chaps. 10, 11, and 16. On the significance of the title "sportsman," see Clive Aslet, *The American Country House* (New Haven: Yale University Press, 1990), pp. 70–78.

9. Herman, *Hunting and the American Imagination*, chaps. 10, 11, and 16. For useful summaries of the symbolism associated with elite hunting, see Gail Bederman, *Manliness and Civilization: A Cultural History of Gender and Race in the United States, 1880–1917* (Chicago: University of Chicago Press, 1995), chaps. 1 and 5; John M. MacKenzie, *The Empire of Nature: Hunting, Conservation, and British Imperialism* (Manchester, England: Manchester University Press, 1988), pp. 7–22.

10. On hunting practices, see Chalmers S. Murray, "Ponemah, 'Happy Hunting Ground,'" *News and Courier*, Apr. 26, 1931, p. A6; Sidney J. Legendre, "Diary of Life at Medway Plantation, Mt. Holly, South Carolina," private collection (copy in

author's possession); Game Books, 1923–1925 and 1934, box 743, Bernard M. Baruch Papers, Seeley G. Mudd Library, Princeton University, Princeton, N.J.; Richard S. Emmet, "Memories of Cheeha Combahee Plantation," *Carologue* (autumn 1999): pp. 18–19. On preferences for certain varieties of game, see Chalmers S. Murray, "Holliday Winters in Peedee," *News and Courier*, Aug. 30, 1931, p. B11; Chalmers S. Murray, "The Wedge on the South Santee," *News and Courier*, Sept. 20, 1931, p. A7; Chlotilde R. Martin, "Belfair – Designed by Artist Owner," *News and Courier*, Jan. 18, 1931, p. B9; Bernard M. Baruch, *Baruch: My Own Story* (New York: Henry Holt and Company, 1957), p. 277.

11. Murray, "Ponemah, 'Happy Hunting Ground.'"

12. Neal Cox, *Neal Cox of Arcadia Plantation: Memoirs of a Renaissance Man* (Georgetown, S.C.: Alice Cox Harrelson, 2003), pp. 35–39, 91–92; "Life Goes to a Party with the Sidney Legendres," *Life* 4, no. 4 (Jan. 24, 1938): pp. 54–57; John J. Seibels to Joseph S. Frelinghuysen, Nov. 10, 1926, box 2, Joseph S. Frelinghuysen Papers, Alexander Library, Rutgers University, New Bunswick, N.J. (hereafter JSF Papers).

13. Herman, *Hunting and the American Imagination*, pp. 156–157.

14. Giltner, *Hunting and Fishing in the New South* (Baltimore: Johns Hopkins University Press, 2008), pp. 78–108.

15. Baruch, *Baruch: My Own Story*, p. 291; Lee G. Brockington, *Plantation Between the Waters: A Brief History of Hobcaw Barony* (Charleston: History Press, 2006), p. 54. The number of African Americans living and working at Hobcaw decreased over time. By the mid-1940s, it fell to about forty. See "Roosevelt Rested at Quiet Manor," *New York Times* (New York, N.Y.), May 8, 1944, p. 36.

16. "Life Goes to a Party," pp. 54, 56–57; "Medway Plantation," *Town and Country* 103, no. 4318 (Mar. 1949): pp. 77–78; Virginia Christian Beach, *Medway* (Charleston: Wyrick and Co., 1999), pp. 57, 81, 84, 86–87, 91, 95, 97.

17. Joseph S. Frelinghuysen to George D. B. Bonbright, May 8, 1925; Agreement between Charles E. Bedeaux and Joseph S. Frelinghuysen, Nov. 23, 1933, p. 4, both in box 1, JSF Papers.

18. Edgar, *South Carolina: A History*, pp. 63–69; Bureau of the Census, *Fifteenth Census of the United States: 1930, Population*, vol. III, pt. 2 (Washington, DC: Government Printing Office, 1931), pp. 784–786.

19. Julia Wood Peterkin, *Roll, Jordan, Roll* (New York: Robert B. Ballou, 1933), p. 13 (quotation) and passim. See also Ambrose E. Gonzales, *The Black Border: Gullah Stories of the Carolina Coast* (Columbia: The State Co., 1922); Reed Smith, *Gullah: Dedicated to the Memory of Ambrose E. Gonzales* (Columbia: University of South Carolina, 1926); Samuel G. Stoney and Gertrude M. Shelby, *Black Genesis: A Chronicle* (New York: Macmillian Co., 1930).

20. See, for example, Coyne Fletcher, "In the Lowlands of South Carolina," *Frank Leslie's Popular Monthly*, Mar. 1891, pp. 280–288; William Henry Johnson scrap-books (3 vols.), SCHS.

21. Emmet, "Memories of Cheeha Combahee Plantation," p. 18; Legendre, *The Time of My Life*, pp. 76–77; Gretta Palmer, "The New Clare Luce," *Look* 11, no. 8 (Apr. 15, 1947): p. 22. On the cultural significance of vigorous exercise, see

David E. Shi, *The Simple Life: Plain Living and High Thinking in American Culture* (New York: Oxford University Press, 1985), pp. 160–164, 175–204.

22. Legendre, *The Time of My Life*, pp. 76–77; Legendre, "Diary of Life at Medway Plantation," entries for May 17, 25, and 28, 1937, and May 23, 1939. For other references to fishing, see Murray, "Ponemah, 'Happy Hunting Ground'"; Train, *A Carolina Plantation Remembered*, p. 103; Brockington, *Plantation Between the Waters*, pp. 53, 94, 106.

23. Chalmers S. Murray, "Metcalf Builds on Peedee Bluff," *News and Courier*, Mar. 22, 1931, p. B14; Chlotilde R. Martin, "New Yorker Reclaims Pimlico," *News and Courier*, May 24, 1931, p. B2; Thomas A. Stone, Boone Hall Scrapbook no. 2, entry for May 25, 1937, SCHS; Legendre, "Diary of Life at Medway Plantation," entries for May 23 and 28, 1937.

24. "Do You Know Your Charleston? Yeaman's Hall," *News and Courier*, Nov. 23, 1931, p. 10.

25. Joseph S. Frelinghuysen to George D. B. Bonbright, May 8 and Nov. 17, 1925, in box 1, JSF Papers; Memorandum, Dec. 18, 1953, fol. 4, box 69, Henry R. Luce Papers, Library of Congress, Washington, D.C. (hereafter HRL Papers).

26. Legendre, "Diary of Life at Medway Plantation," entries for Nov. 22, 1937 and fall 1939.

27. Legendre, *The Time of My Life*, p. 70.

28. "Do You Know Your Charleston? Dominick Estate," *News and Courier*, Dec. 5, 1932, p. 10; Chlotilde R. Martin, "Dominick Builds Among Giant Oaks," *News and Courier*, Feb. 8, 1931, p. B14; J. V. N., Jr., "Do You Know Your Low-country? Prospect Hill," *News and Courier*, May 9, 1938, p. 10; Chlotilde R. Martin, "The Hutton Estate of 16,000 Acres," *News and Courier*, Jan. 25, 1931, p. B1; Linder, *Historical Atlas of the Rice Plantations of the ACE River Basin*, p. 138; "New York Broker Builds at Delta," *News and Courier*, Apr. 19, 1929, p. 12; "Polo Player Buys Georgetown Estate," *News and Courier*, May 4, 1931, p. 3.

29. Frelinghuysen to Bonbright, Nov. 17, 1935, box 1, JSF Papers; Henry R. Luce to Harold J. T. Horan, Feb. 5, 1938, fol. 7, box 98, HRL Papers.

30. Mark Rothery, "The Shooting Party: The Associational Cultures of Rural and Urban Elites in the Late Nineteenth and Early Twentieth Centuries," in *Our Hunting Fathers: Field Sports in England After 1850*, ed. R. W. Hoyle (Lancaster, England: Carnegie Publishing, 2007): pp. 96–118.

31. Vivian, "The Leisure Plantations of the South Carolina Lowcountry," pp. 140–141.

32. Legendre, "Diary of Life at Medway Plantation," entry for Mar. 29, 1940.

33. See, for example, Murray, "Attracted by Climate"; Martin, "Low-Country Plantations Stir"; "Northern Citizens Now Returning to their Homes Here," *George-town Times*, Dec. 20, 1929, p. 6; "Hunting Grounds Echo Gun Shots," *News and Courier*, Dec. 9, 1930, p. 10; "Arrivals at Charleston," *New York Times*, Nov. 17, 1932, p. 16; "Winter Residents Reopen Estates," *News and Courier*, Nov. 25, 1938, p. B4.

34. "To Save Carolina Game," *New York Times*, Mar. 25, 1932, p. 7; "Plantation Group Meets at Wedge," *News and Courier*, Mar. 25, 1934, p. A14.

35. "Plantation Group Elects Officers," *News and Courier*, Jan. 6, 1935, p. 10; "Plantation Men Meet at Lunch," *News and Courier*, Mar. 3, 1935, p. B4; "Plantation Men Whitney Guests," *News and Courier*, Feb. 24, 1936, p. 3; "Plantation Men Meet at Lunch," *News and Courier*, Jan. 11, 1937, p. 10.

36. "Plantation Group Meets at Wedge"; "Plantation Men Meet at Lunch," *News and Courier*, Mar. 3, 1935, p. B4. See also "Plantation Men Whitney Guests" and "Plantation Group Elects Officers."

37. "Plantation Group Meets at Wedge"; "Plantation Group Plans for Dinner," *News and Courier*, Dec. 13, 1935, p. A12; "Plantation Men Meet at Lunch," *News and Courier*, Mar. 3, 1935, p. B4; "Plantation Group Elects Officers"; "Plantation Men Whitney Guests"; "Plantation Men Meet at Lunch," *News and Courier*, Jan. 11, 1937, p. 10; David Doar, *Rice and Rice Planting in the South Carolina Low Country* (Charleston: Charleston Museum, 1936), pp. 69–70.

38. Henry P. Davis, "High Dogs of the Low Country," *Country Life* 77, no. 5 (Mar. 1940): pp. 82–83; Chlotilde R. Martin, "Low Country Gossip," *News and Courier*, Feb. 21, 1932, p. A11.

39. Davis, "High Dogs of the Low Country," pp. 81–83.

40. On field trials, see Bernard Waters, *Training the Hunting Dog for the Field and Field Trials* (New York: Forest and Stream Publishing Co., 1901), chaps. 19 and 20; Vinton P. Breese, "Show and Field Dogs," *Town and Country* 90, no. 4146 (Feb. 15, 1935): pp. 60–61, 65; Peter R. A. Moxon, *Gundogs: Training and Field Trials*, 18th ed. (Shrewsbury, England: Quiller, 2010), chap. 9.

41. Davis, "High Dogs of the Low Country," pp. 82–83.

42. Ibid., pp. 81–83.

43. Legendre, *The Time of My Life*, p. 76.

44. Flora M. C. Stone to My Dear Ones, Boone Hall Scrapbook no. 2, entry for Mar. 5, 1937, SCHS. Stone's comments are strikingly similar to those of the eminent historian Ulrich B. Phillips, who characterized social conditions at a U.S. Army training installation in Georgia during World War I as having "a plantation atmosphere." He attributed it to a combination of white authority and black subordination. See Ulrich B. Phillips, *American Negro Slavery* (New York: D. Appleton and Co., 1918), pp. viii–ix.

45. Claire Luce, "The Victorious South," *Vogue*, June 1, 1937, p. 120.

46. Luce, "The Victorious South," pp. 121–122.

47. Legendre, "Diary of Life at Medway Plantation," entry for fall 1940.

48. Ibid., entry for Feb. 26, 1940.

49. Ibid., entry for Mar. 9, 1940.

50. Joseph S. Frelinghuysen to Herman Hansen, Apr. 30 and May 4, 1931, box 6, JSF Papers.

51. "Medway Plantation," *Town and Country* 103, no. 4318 (Mar. 1949): pp. 76–79; Stephanie E. Yuhl, *A Golden Haze of Memory*, p. 143. Other scholars have also noted African Americans' role in performing for white hunters as part of what the latter considered a "complete sporting experience." See, for example, Giltner, *Hunting and Fishing in the New South*, p. 94.

52. Yuhl, *A Golden Haze of Memory*, pp. 127–154.

53. The names of several plantation owners and other northerners appear in a sales ledger of copies for *The Carolina Low-Country*. See Account Book, 1931–1933, fol. 14, box 715, Records of the Society for the Preservation of Spirituals, SCHS.

54. Lydia Parrish, "A Heritage We Must Not Lose," *Country Life* 69, no. 2 (Dec. 1935): pp. 50–55, 62.

55. "Plantation Singers by Moonlight at Cypress Gardens," *News and Courier*, Apr. 19, 1935, p. A14.

56. "The Oaks, A Restored Mansion of the South," *Country Life* 29, no. 2 (Dec. 1915): p. 54.

57. Emmet, "Memories of Cheeha Combahee Plantation," p. 18.

58. Laura C. Hemingway, "Williamsburg Boasts Many Game Preserves for Hunt," *News and Courier*, Jan. 11, 1937, p. 3.

59. For a general description of Christmas at lowcountry plantations, see Doar, *Rice and Rice Planting in the South Carolina Low Country*, p. 33.

60. "Ol' Maussa Greenway Gives Party for Plantation Hands," *News and Courier*, Jan. 10, 1935, p. 3.

61. Thomas A. Stone Journal, entries for May 17 and 19, 1937, in Boone Hall Scrapbook no. 3, SCHS. A somewhat similar birthday celebration took place at Prospect Hill Plantation in January 1934. Franklyn L. Hutton and his wife threw a lavish celebration for the twenty-first birthday of their son, Woolworth Donahue. According to the *News and Courier*, while the festivities were underway, "the plantation negroes came to a window outside the dining room and sang 'Happy Birthday to You.'" The Jenkins Orphanage Band played, and Lieutenant Joseph F. Wise of the Charleston Police staged an arrest of Mr. Donahue for "violating hunting laws." According to the newspaper, this was "one of the stunts of the evening." See "F. L. Huttons Give 21st Birthday Party for Woolworth Donahue," *News and Courier*, Jan. 11, 1934, p. 5.

62. Chalmers S. Murray, "Arcadia, Where LaFayette Stopped," *News and Courier*, June 21, 1931, p. B5; Cox, *Neal Cox of Arcadia Plantation*, pp. 86–88; "Mepkin Plantation, Moncks Corners, S.C.," *Architectural Forum* 66, no. 6 (June 1937): p. 522; Brockington, *Plantation Between the Waters*, p. 55; V. C. Barringer to F. A. Collins, Jan. 12, 1942, fol. "Barringer, Victor C., 1942–43," box 133, Harry F. Guggenheim Papers, Library of Congress, Washington, D.C. (hereafter HFG Papers).

63. Suzanne Cameron Linder and Marta Leslie Thacker, *Historical Atlas of the Rice Plantations of Georgetown County and the Santee River* (Columbia: South Carolina Department of Archives and History for the Historic Ricefields Association, Inc., 2001), p. 144; J. V. Nielsen, Jr., "Harry Guggenheim Saves St. Thomas's Church," *News and Courier*, Nov. 8, 1936, p. C3; J. S. Frelinghuysen to George D. B. Bonbright, Oct. 26, 1925, box 1, JSF Papers.

64. Claire Boothe Luce to Franklin Delano Roosevelt, Jan. 9, 1937, quoted in Sylvia Jukes Morris, *Rage for Fame: The Ascent of Claire Boothe Luce* (New York: Random House, 1997), p. 299; W. B. Seabrook to Thomas A. Stone, Sept. 30, 1938, Boone Hall Scrapbook no. 3, SCHS. On the New Deal in the lowcountry, see Jack Irby Hayes, *South Carolina and the New Deal* (Columbia: University of South Carolina Press, 2001), chaps. 3–5 and 11; Fraser, *Charleston! Charleston!* pp. 380–383.

65. "The Yankee Doodle Salon," *Fortune*, Dec. 1937, pp. 123–127. On café society, see also Cleveland Armory, *Who Killed Society?* (New York: Harper and Brothers, 1960), pp. 20, 107–108; Frederic Cople Jaher, *The Urban Establishment: Upper Strata in Boston, New York, Charleston, Chicago, and Los Angeles* (Urbana, Ill: University of Illinois Press, 1982), p. 280; Thierry Courdert, *Café Society: Socialites, Patrons, and Artists, 1920–1960* (Paris: Flammarion, 2010). On changes in upper-class culture between the world wars, see Frederic Cople Jaher, "The Gilded Elite: American Multimillionaires, 1865 to the Present," in *Wealth and the Wealthy in the Modern World*, ed. W. D. Rubinstein (New York: St. Martin's Press, 1980), pp. 201–202.

66. Legendre, "Diary of Life at Medway Plantation," entry for Apr. 17, 1941. On labor shortages, see Alexander Hehmeyer to H. R. Luce, Nov. 5, 1942, fol. 11, box 85, Claire Boothe Luce Papers, Library of Congress, Washington, D.C.; V. C. Barringer to Harry F. Guggenheim, Sept. 6, 1943, fol. "Barringer, Victor C., 1942–43," box 133, HFG Papers; Gertrude S. Legendre, *The Sands Ceased to Run* (New York: William-Frederick Press, 1947), pp. 4–5; Cox, *Neal Cox of Arcadia Plantation*, p. 111. On wartime activity, see Fraser, *Charleston! Charleston!*, pp. 387–393; Fritz P. Hamer, *Charleston Reborn: A Southern City, its Navy Yard, and World War II* (Charleston: History Press, 2005), pp. 39; Stokes, *Myrtle Beach*, pp. 50–581; Edgar, *South Carolina: A History*, pp. 510–515; Brockington, *Plantation Between the Waters*, pp. 108–109.

67. Robert Sullivan, "The Yankee Invasion of S. Carolina Ebbs," *New York Daily News* (New York, N.Y.), May 21, 1944, pp. 44, 47; Linder and Thacker, *Historical Atlas of the Rice Plantations of Georgetown County and the Santee River*, p. 514; Alberta Morel Lachicotte, *Georgetown Rice Plantations* (Columbia: The State Printing Co., 1955), pp. 77, 101, 131.

68. On these developments, see especially Bederman, *Manliness and Civilization*; Michael Kimmel, *Manhood in America: A Cultural History* (New York: Free Press, 1996), chaps. 3–5; E. Anthony Rotundo, *American Manhood: Transformations in Masculinity from the Revolution to the Modern Era* (New York: Basic Books, 1993), chaps. 8–10.

69. Jaher, "The Gilded Elite," pp. 257–263; Pitirim Sorokin, "American Millionaires and Multi-Millionaires: A Comparative Statistical Study," *Journal of Social Forces* 3, no. 4 (May 1925): pp. 639.

Chapter 2

Reviving and Restoring Southern Ruins

Reshaping Plantation Architecture and Landscapes in Georgetown County, South Carolina

Jennifer Betsworth

In December 1937, the *Georgetown Times* boasted "more wealthy Northern-ers maintain winter homes in this county than in any other section of the South of similar area."[1] During the preceding fifty years, wealthy outsiders had purchased the majority of Georgetown County plantations and turned them into showplaces. In 1930, only seven of thirty-five surviving planta-tions from the colonial and antebellum eras remained in local hands. In the decades after the Civil War, the South Carolina lowcountry became a land of "deserted fields growing up in forest, of ragged dying gardens and grim, cold, pathetic houses, solemnly awaiting their doom by fire or dilapidation." Wealthy outsiders seeking winter homes saved many plantations from com-plete desolation. The newcomers rehabilitated old houses, brought fields back into production, and spurred on local economies through the vast amounts of money they spent. Lowcountry author and historian Samuel Gaillard Stoney attributed the "rediscovery" of the region to cars and good roads.[2] In fact, the movement—called the "Second Yankee Invasion" by some—owed more to the passenger train than the personal automobile.

President Grover Cleveland's well-publicized visits to former Confederate General E. P. Alexander's South Island preserve in the 1890s helped adver-tise the lowcountry's excellent duck hunting.[3] Northern business partners and friends purchased South Carolina plantations for use as game preserves and established hunting clubs. In Georgetown County, the earliest and most prominent of such clubs was the Santee Gun Club. Formed in 1898 by men from Boston and New York, it quickly took control of 20,000 acres in the Santee Delta, immediately below Georgetown.[4] Hunting clubs offered opportunities for "the city-bound man [to obtain] a vacation 'in the woods'" and introduced northern businessmen to South Carolina.[5] As northerners became familiar with the region, some purchased plantations to use as private retreats and winter residences. The "Second Yankee Invasion" occurred throughout the entire South but took different forms in different regions. In Georgetown County it attracted people who desired privacy, high-quality hunting conditions, and a landscape imbued with a sense of history.[6]

At its height during the late antebellum era, Georgetown County, north of Charleston, figured among the wealthiest rice-growing regions in the United States. By 1840, slaves on Georgetown plantations produced almost half of all rice grown in the country. Planters built large, stylish houses to express their wealth and status. After the Civil War, attempts to mechanize rice production failed. Between 1893 and 1911, a series of hurricanes crippled the local rice economy. As commercial production ended, cash-strapped families viewed the houses and land of once-productive plantations as saleable assets. Timber companies and local interests purchased some, but northerners who appreciated the beauty of the land and saw potential for hunting reserves soon became the most common buyers.[7]

In Georgetown County, the remaking of old plantations for leisured use began somewhat earlier than elsewhere on the South Carolina coast. At the turn of the twentieth century, Bernard Baruch, the financier and presidential advisor, and Isaac Emerson, the Bromo-Seltzer millionaire, each bought and consolidated several well-known plantations. Purchases became more frequent during the 1920s and 1930s, when men such as Tom Yawkey, the owner of the Boston Red Sox, and Archer Huntington, arts patron and philanthropist, bought land and created large estates. Northerners modified plantation buildings and landscapes according to their tastes and intended uses. Locals followed these changes with interest. Despite the attention these developments received from contemporaries, however, historians have paid them limited attention. A mélange of old and new, plantations remade as winter estates defy simple categorization. Their origins and development remain largely unexplored.[8]

Collectively, the new owners launched a building and renovation campaign of sizeable proportions. They transformed creaky houses, deteriorating outbuildings, and overgrown landscapes. Most sought to amplify or create anew the romantic atmosphere popularly associated with southern plantations. Virtually to a rule, northerners allowed elements of former working landscapes and, by implication, their association with slavery, to remain glossed over by the patina of time. Many drew inspiration from the Colonial Revival, which championed visions of a pure and simple past and informed renovations of plantation buildings and the design of new houses and landscapes. Owners and architects expressed appreciation for old buildings and historical styles but lacked consistent terminology in describing their work, let alone a standard methodology. Most valued authenticity and original details but saw them as malleable, depending on the circumstances. Commentators tended to credit northerners with "restoring" old plantation houses and recognized their actions as preventing further decay, but the language they used told little about the treatments employed. Once architects, builders, and landscapers had finished their work, owners and their guests reveled in elegant surroundings that bore only passing resemblance to the large-scale rice plantations they replaced. Locals paid close attention to the revitalization of the county's grand old homes and, although appreciative of northerners' efforts, recognized the newcomers as challenging established social hierarchies. Careful study of the process that unfolded illuminates the creation of grand "plantations" devoted to leisured use.

WINTERING AMID AN IDEALIZED SOUTHERN LANDSCAPE

Southern cities and health resorts became popular destinations before the Civil War. In February 1855, Frederick Law Olmsted noted South Carolina public-houses "over-crowded with Northerners" escaping winter for a healthier, warmer climate.[9] Doctors encouraged patients to go to southern hot springs, piney woods, and beaches to cure respiratory illnesses and nervousness brought on by the conditions of "modern" life. Patients responded in large numbers.[10] By the early 1890s, northern capitalists began traveling to the South for leisure. Journeys to southern winter resorts, usually in Florida, soon became an important part of the social seasonal round. "So great is our own country that a man may in two days move from winter to summer, from March to July," exclaimed the editor of the *Chautauquan*. He added, "who cares if these people do leave a million [dollars] or two in Florida every winter? Florida does not object."[11]

About the same time, wealthy individuals turned to the South Carolina lowcountry as a new vacation destination and desirable location for winter homes. When northerners looked south for a vacation, their perception was shaped by a century of oppositional regional identity formation. The images of a fundamentally agrarian South and an industrial North traced their roots to the early nineteenth century. Postwar Lost Cause ideology built upon these images to celebrate not only the rightness of the Confederate cause but the elite southern lifestyle that the end of slavery brought to a close. Railroad advertisements and popular literature directly referenced iconic ideals of southern culture by noting the grandeur and hospitality that awaited visitors.[12]

Although the South lost the Civil War, historians have argued that southerners won the peace through their success in shaping memory of the conflict.[13] The United Daughters of the Confederacy, which was created from localized memorial associations, ardently supported the Lost Cause. The group sought to vindicate and craft the memory of the Confederate past through monument-building campaigns and the protection and dissemination of the "true" history of the South.[14] A reunion culture developed in national literature, which often provided a romantic, sentimental image of the South as well as the conflict and causes of the war itself. Tales of romances between northern men and southern women were particularly popular by the end of the century, whether they reported actual events or were fictional.[15] Although sentimentalization of the conflict helped to strengthen calls for reconciliation, white writers typically ignored slavery as the cause of the war or presented mythologized images of faithful blacks who remained on their plantations.[16]

Travelogues, typically presented as short articles in national magazines, reinforced the image of the South as offering opportunities for visitors to encounter disappearing but still-tangible pasts. Coyne Fletcher's "In the Lowlands of South Carolina," published in 1891, blends an idealized account of Georgetown's past with the experience of traveling into its harbor, through the town, and past old plantations. Fletcher's descriptions hint that he felt he was not only passing by old churches and houses, which he describes as the "heritage of the people of the South," but that he was visiting with the past itself. He laments the decay of old mansions, mutilated by vandals, and the "desolate expanse" of former rice fields, but otherwise steadfastly avoids contemporary descriptions of local conditions and culture.[17]

The traditional southern pastoral ideal balanced with the decline of the rice industry created landscapes that seemed familiar and looked increasingly feral. Overgrown rice fields and timber tracts alongside spreading live oaks covered in Spanish moss created a sense of mystery and ruin amid rustic surroundings. A 1932 issue of *Country Life* described the lowcountry as a place where one could find "hospitality, riding to hounds, and all the other arts of gentlemen. There is no part of America more remote from pressure salesmanship and its philosophy of hurry and shove; there is no place where the sound of a stock ticker would seem so ill-suited to the surroundings." Instead of escaping to primitive nature, northerners in the lowcountry could travel back to the more recent "antimodern refuge" of the mythologized Old South of "soft voices and good manners."[18]

The mythology and memory of the Old South inspired northern interest in plantation landscapes. Visitors tended to see decaying plantations as romantic ruins more than working landscapes rendered dormant by cataclysmic upheavals. Although some northerners showed passing interest in the large numbers of African Americans who remained present, few noted the land's direct connection to slavery. Northerners' acquisition made preservation of plantation landscapes and buildings possible. By the end of the 1920s, many observers credited wealthy sportsmen and sportswomen with saving grand houses, plantation graveyards, old trees, and other "remnants of culture of a day gone by."[19]

Individual owners sought to secure beautiful, secluded estates for their use. The broad pattern, however—the "Second Yankee Invasion" as a whole—played a crucial role in the formation of historical memory. Through physical revitalization of plantation buildings, new owners contributed to a popular reframing of their historic use. Northerners' activities made former agricultural enterprises into sites of leisure and settings for parties, relaxation, and rejuvenation, albeit unevenly. Northerners' efforts focused mainly on big houses and adjoining landscapes. Casual interest in former agricultural buildings and rice fields rarely led to extensive rehabilitation. Planters' dwellings received refurbishment, new additions, and newly beautified surroundings while portions of the historical built environment related to labor and agricultural enterprise slipped away, left to continue their decay.

Buyers of South Carolina plantations participated in the American country house movement. Between 1880 and 1930, wealthy Americans built country houses near cities such as Philadelphia, Boston, and New York to serve as weekend retreats. The country house movement paralleled the nineteenth-century creation of country estates in Britain and responded to rapid urbanization and the concerns about declining living conditions that accompanied it.[20] Wealthy families built sumptuous mansions on Long Island, in the Berkshire Mountains of western Massachusetts, and on the outskirts of cities such as Philadelphia and Wilmington. Although the form and style of country estates varied, most had stately dwellings, formal gardens and landscaping, outbuildings, and enough land to suggest self-sufficiency. Through the design, style, and use of such estates, owners projected status, aspirations, and identity.[21]

Southern country homes offered distance and respite from popular resorts. Despite their removal from the major centers of upper-class life, owners cultivated an image of rural gentility. Rather than building grand new houses, many chose to restore surviving colonial and antebellum plantation buildings. Large-scale restoration projects quickly became common. The *Georgetown Times* reported on examples into the 1940s.[22] When northerners built new houses, most chose the Colonial Revival style.[23] Often, new owners retained the historic names and boundaries associated with their lands. The resulting combination of changes made it difficult to see where old plantation buildings and landscapes ended and new estates began.

Although historians have emphasized the appeal of "ready-made" estates, new owners consciously chose to restore or build in a form consistent with the image of the Old South. This decision reflected how they wanted to be perceived by their northern friends as well as their new southern communities. Former plantation land offered northerners the unique opportunity to buy into the image of the southern aristocracy and become an immediate part of the local social hierarchy. While a visitor could experience southern hospitality and admire the grandeur of a plantation, only the owner of such a property would benefit from the class-based status of owning a grand piece of southern land.[24]

RESTORING GEORGETOWN COUNTY PLANTATIONS

Years of deferred maintenance took a toll on most Georgetown County plantations by the beginning of the twentieth century. Moldering buildings may have inspired poetic visions and interested passing visitors, but to new owners they posed significant problems. Although not all of the new owners chose to rehabilitate existing structures, many called upon local craftsmen and architects to reclaim some of their former glory.

The problem of how to treat old buildings became a focus of attention for architects, designers, and architectural writers during the early twentieth century. Enthusiasm for saving and restoring the homes of great men across the country grew during the nineteenth century. After learning that George Washington's home, Mount Vernon, was threatened by development, the Mount Vernon Ladies Association formed in 1853. They purchased the property and began a restoration after five years of dedicated fundraising and campaigning. Similarly, a group coalesced around the Paul Revere house in Boston at the turn of the twentieth century. They saved the house from demolition and began an extensive restoration in 1907. By contrast, appropriate treatments for the thousands of less-pedigreed old buildings along the eastern seaboard remained undetermined. Architects, landscape architects, antiquarians, and self-reliant carpenters debated proper methods of modifying old buildings for new use. Early advice manuals encouraged substantially modifying the style and plan of old homes to essentially create something new. By the turn of the twentieth century, however, authors began to stress the value of historic details and authenticity, an emphasis that became more entrenched in later decades. A clear professional consensus, however, remained elusive.[25]

The men and women who purchased Georgetown County plantations did not proclaim associations with the nascent preservation movement. Most revitalized and renovated former plantation homes in ways that suited their needs and whims. The decision to live in an old house balanced romantic and pragmatic sensibilities. On the one hand, it avoided the need to build a new dwelling. On the other, it supplied a direct connection to what many Americans of the era viewed as the only true aristocracy in the nation's history. In most cases, neither contemporary sources nor new owners explained what led the latter to make the choices they made and or the underlying rationale. In most cases, northerners' aims and ambitions remain opaque. Isaac Emerson and Cornelia Sage offer two examples. They figured among the first northerners to "preserve" pre-Civil War plantation houses in Georgetown County.

Isaac Emerson, a native North Carolinian, invented Bromo-Seltzer, a popular antacid, in Baltimore in 1888. The product's widespread success soon made him a millionaire. In 1906, Emerson made headlines in Georgetown when he purchased Prospect Hill Plantation for use as a winter home. Residents of the area treasured the plantation's colonial-era main house. According to local lore, the Marquis de LaFayette slept in it immediately following his arrival in the colonies during the American Revolution. Although the *Georgetown Times* initially celebrated Emerson's purchase simply by touting the benefits of turning "valueless" old plantations into "good hard Yankee cash," popular sentiments apparently prized Prospect Hill for other reasons. The *Times* made this apparent a few weeks later when it excitedly reported that Emerson had decided to restore the house and had engaged a number of well-paid expert workmen to undertake the project.[26]

Under Emerson's direction, restoration involved more than refurbishing and ensuring structural integrity. He also added modern conveniences; built a pair of symmetrical wings that contained a ballroom and a gymnasium; and added a number of outbuildings. Consistent with the restoration ethic of the

Figure 2.1 Circa 1900 postcard showing main house at Prospect Hill Planation. Pharmaceutical tycoon Isaac Emerson purchased Prospect Hill and several other plantations to create a sprawling estate and named it "Arcadia." Like most northern buyers, he renovated and enlarged the house at Prospect Hill to make it suitable for hosting guests. *Source*: Courtesy Morgan and Trenholm Collection, Georgetown County Library, Georgetown, S.C.

day, these additions mirrored the design of the original dwelling. Their size and placement, however, made the building significantly more impressive. Emerson's modifications underscored a desire to create an estate suited for entertaining and leisure. He returned the house to its "highly cultivated state" not by striving for authenticity but, rather, by making it a setting for upper-class activity once again.[27]

Following his purchase of Prospect Hill, Emerson bought several adjoining tracts of land, some of which encompassed the remnants of antebellum plantations. He named his estate Arcadia in reference to the idyllic pastoral utopia of Greek mythology.[28] Emerson's choice demonstrated personal alignment with the growing national interest in nature and the intentions of English country house owners who strove to meet the unattainable Arcadian myth at their own estates.[29] Emerson must have enjoyed the process of restoration. Ten years after acquiring Arcadia he purchased and restored Brooklandwood, a Baltimore mansion built during the 1790s. Between Arcadia, Brookland-wood, and Sagamore, his daughter Margaret Emerson's summer home in the Adirondacks, Isaac Emerson maintained a full schedule of social engagements and heavy estate management responsibilities.[30]

Henry M. and Cornelia Sage of New York carried out the most unusual estate-making efforts in Georgetown County. Henry Sage, the heir to his father's Albany lumber business, avidly collected antiques and traveled widely. In 1920, he and his wife renovated their 1890s Shingle-style house in Menands, New York, into a Georgian Revival mansion complex.[31] When they first traveled to Georgetown, their recent architectural foray was fresh in their minds. In the winter of 1929, the couple fell in love with the beautiful public gardens at Belle Isle plantation. The owner, Frank E. Johnstone, agreed to lease the plantation to them for ten years, provided he could continue charging admission to his gardens. Although they loved the setting, the Sages demanded as part of the lease that Johnstone dismantle the late nineteenth-century house on the property. The couple had determined that renovating the house to fit their image of a grand plantation mansion would be too costly, so, in what may have been the first move of its kind in South Carolina, they decided to relocate a 100-year-old house from Newberry, South Carolina, to Belle Isle.[32]

Built in 1829, the Mendenhall house met the Sages' expectations for high-style architecture. The Federal-style residence had a porch supported by hand-carved columns and interiors filled with detailed woodwork. A mahogany stair dominated the entry hall, and original wallpaper covered the walls of several rooms on the first floor. When workers dismantled the house in preparation for the move, Cornelia had the wallpaper sent to New York for conservation and repair. She spared no expense in ensuring that all pieces of the structure survived the journey to Belle Isle and received careful restoration upon reaching the site.[33] At the end of the ten-year lease, Cornelia Sage offered to purchase Belle Isle, but Frank Johnstone refused her offer. Johnstone had suffered financially during the Great Depression and planned to move his family into the Mendenhall house. Before she left, Cornelia removed the wallpaper and staircase, a right she had carefully stipulated in the lease.[34]

While she lived at Belle Isle, Cornelia continued exploring other plantations. In 1931 she purchased Dover Plantation, directly south of Belle Isle. She discovered an abandoned, deteriorating house at the nearby Woodlawn Plantation and in 1940 convinced the owner to sell it to her. After her lease at Belle Isle ended, Cornelia moved the Woodlawn house to Dover plantation and undertook a full-scale restoration. She had the wallpaper and staircase from the Mendenhall house installed and also added a salvaged doorway, a balcony, and a Palladian window from a house in Savannah. At the time, combining elements from different houses rarely drew censure and sometimes inspired praise. In her 1955 book *Georgetown Rice Plantations*, Alberta Lachicotte commended the new house as beautifully done and "unsurpassed by any plantation home in the state."[35]

Restoration often dramatically changed Georgetown County plantation houses. Northerners preserved old buildings while adding modern conveniences, salvaged elements, and massive additions. Often architecturally invasive, their efforts damaged historic materials in order to create comfortable and commodious residences. Moreover, interest in sumptuous architecture rarely extended to agricultural buildings. Although owners made use of surviving stables and barns, their needs paled in comparison to those of plantations operated as commercial enterprises.

These case studies offer a snapshot of how personal preferences and professional expertise shaped the material development of twentieth-century plantations. Although tracing the influence of national discussions about proper restoration practices with any degree of precision is virtually impossible, new owners and the builders and architects they hired clearly worked with knowledge of contemporary debates. The renovations and restorations that northerners' carried out yielded twentieth-century mansions more than colonial plantation dwellings. By making extensive changes, new owners wove themselves into the physical history of Georgetown-area buildings and landscapes.

NEW COLONIAL PLANTATIONS AND LANDSCAPES

Though restoration proved exceedingly popular, northerners often built new houses on the lands they purchased. Wealthy individuals associated rustic buildings and bungalows with wilderness vacations and the rage for outdoor recreation that captivated middle- and upper-class Americans of the early twentieth century. Seeing rustic styling as appropriate to virtually any naturalistic setting, northerners brought it with them to South Carolina, where they used it mainly for hunting clubs and camps. Gravel Hill Plantation is a notable exception. In the early 1910s, R. P. Huntington, a retired architect, built a rustic-style complex on his 2,500-acre plantation in Hampton County. Modeled on his summer home in the Adirondacks, the complex included a rambling house with deep overhanging eaves, several cabins, a kennel, and a barn.[36] Similarly, the Kinloch Gun Club initially planned to build an architect-designed rustic log clubhouse with bark-edge clapboards.[37] Otherwise, exotic and eclectic designs proved rare. Archer and Anna Hyatt Huntington's Atalaya, a palatial residence modeled after a Spanish ruin, is unique among the houses that wealthy northerners built in Georgetown County.[38]

The majority of new homes built by northerners featured Colonial Revival styling, as if to appear that they belonged to the plantation past. As Augusta Patterson stated, "A Colonial house to be successful must be a picture as well as a building."[39] What could be more fitting for a region that prided itself on its romantic, picturesque image? Just as in popular literature about antebellum life, architects and their clients relied on their imaginations more than historical facts in creating impressive new houses suited to contemporary tastes.

The Colonial Revival transformed beliefs about the treatment of old houses. Spurred on by the 1876 Centennial celebration and apprehensions about the dilution of American values in an era of massive European immigration, architects, designers, and architectural writers saw the Colonial style as the only distinctively American mode of design.[40] Millions seized on it as a vehicle for expressing patriotism, respectability, and refined taste. Middle-class Americans especially favored Colonial Revival designs. Tastefully-styled houses filled with new colonial-style furnishings became the new vogue. Through the Colonial Revival and the contemporaneous Arts and Crafts and Beaux-Arts styles, Americans showed preferences for simplicity and tradition over the ornateness and extravagance of Victorian design. The combination of nationalistic sentiment and growing concern about the changes caused by industrialization inspired efforts to preserve colonial-era buildings, collect antiques, and write regional histories.[41] The southern form of the Colonial Revival did not come into its own for several decades. White-columned facades became its hallmark by the turn of the century, and, over time, columnar mansions set in verdant surroundings became a sort of architectural shorthand for commemoration of the Old South.[42]

Whether in newspaper articles, advertisements, or travel books, commentators consistently praised the South as a haven from the ills and rush of modernity. Northerners' preferences for traditional styles complimented popular views of the region. Few could—or wanted to—imagine the jarring effect of Modern houses and outbuildings in a region of romantic old plantations. In many cases, the designers of new Georgetown County plantation houses are elusive. Contemporary accounts did not consistently identify them by name. South Carolina architects received commissions for about half the houses that northerners built, while prominent northern firms designed the others.[43] Irrespective of background, all satisfied clients' desires by designing modern dwellings that appeared to have stood for centuries. Houses erected at Hobcaw Barony and at Friendfield and Wedgefield plantations provide representative examples.

In 1904, an advertisement for "Hobcaw Barony" appeared in the October edition of *Country Life in America*.[44] Within a few months, Bernard Baruch purchased the plantation. By 1907, he also purchased nine other plantations on the Waccamaw Peninsula, thereby reassembling the entirety of the original Hobcaw Barony, a 12,000-acre colonial land grant. After making a fortune as a financier in New York City, Baruch eagerly seized the opportunity to create a winter residence in his home state.[45]

For years, Baruch and his family occupied a two-story Queen Anne-style dwelling that they playfully called the "Old Relick." Built about 1890, the building supplied the Baruchs with comfortable accommodations during their

Figure 2.2   Bernard M. Baruch and his family occupied this Queen Anne-style residence during their stays at Hobcaw Barony from 1907 until it burned in December 1929. Built circa 1893, it illustrates the absence of pretention that prevailed in the era before wealthy sportsmen and sportswomen turned old plantations into handsome estates. The Baruchs lovingly referred to the dwelling as the "Old Relick." *Source:* Courtesy Belle W. Baruch Foundation, Georgetown, S.C.

visits to the Barony. In December 1929 the house caught fire in the midst of their holiday celebrations and could not be saved. The impressive fire could be seen in Georgetown, across the harbor. Baruch made plans to build a new house immediately. An early, exaggerated report claimed that a fireproof, twenty-room "three story colonial mansion" of brick, steel and concrete would soon take the place of the old home.[46] Baruch selected the Columbia architectural firm of Lafaye and Lafaye to design the new building. Known mainly for designing public buildings, Lafaye and Lafaye had recently completed a hospital in Camden, South Carolina, that Baruch financed. With a total cost of around $100,000, the Hobcaw house became the largest residential project undertaken by the firm to date.[47]

Lafaye and Lafaye adroitly addressed Baruch's concerns regarding fire and aesthetics by creating a Colonial Revival home from unabashedly modern materials: brick veneer on a steel frame. Although the new house was built near the site of the original home, it was an entirely new design, not a replica. A massive portico with square columns gave the façade an appearance similar to Mount Vernon and added a measure of grandiosity. Baruch saw the new house and the surrounding grove of magnolias and live oaks as creating an appropriate atmosphere for the Barony. He reported that while admiring the home and its surroundings, Otto Kahn, a New York banker, stated, "I really know for the first time why Southerners feel about the South as they do."[48]

In 1930, Radcliffe Cheston, Jr., an investment banker from Philadelphia, Pennsylvania, purchased Friendfield Plantation on the Sampit River. Cheston had an affinity, or at least an appreciation, for old houses. His family made their home in Emlen House, a stone dwelling in Oreland, Pennsylvania, built largely before 1730. The original eighteenth-century house at Friendfield had burned in 1926. Cheston decided to build a new house on the old foundations. To take on this task, he hired the prominent Philadelphia firm of Mellor and Meigs, which had designed the family's summer home in Maine. Cheston's house became the only project the firm carried out in the South. In keeping with their established reputation, they successfully adapted regional architectural styles for the new building.[49]

By examining the old foundation and chimney and historic paintings of the house, Mellor and Meigs determined its original proportions. The new house was finished in 1932. The general massing and roofline is similar to the old Friendfield house, but a dramatic full-height pedimented portico gives the façade a monumental appearance. Additional cornice detail also makes the house appear more grandiose. After several years of driving and parking on the front lawn, the Chestons demanded a more refined solution. They brought Mellor and Meigs back to Georgetown in 1936 to add an Italianate loggia attached to a car park and a new primary entrance on the west side of the house.[50]

Figure 2.3 Following the fire that destroyed the Old Relick, Baruch erected this handsome Colonial Revival mansion. Designed by Lafaye and Lafaye of Columbia, the building features fireproof construction. *Source*: Courtesy Belle W. Baruch Foundation, Georgetown, S.C.

Although Wedgefield Plantation changed hands no less than six times during the early twentieth century, its original colonial-era house remained largely intact. When Robert Walton Goelet purchased the plantation in 1935, he promptly had it torn down. Goelet, a financier and heir to a New York real estate fortune, was born into a tradition of vacation homes and architectural patronage. Goelet's father, Ogden, hired Richard Morris Hunt to design Ochre Court, a chateauesque mansion in Newport, Rhode Island, in 1892. Ogden's brother, Robert, commissioned McKim, Mead, and White to build a Shingle-style house nearby three years later. Only three months after purchasing Wedgefield, Robert Walton Goelet hired William L. Bottomley, a prominent New York architect popular for his romantic, historicist designs, to build a new house at Wedgefield.[51]

The house that Bottomley designed is a showplace. The long, one-story brick residence differs from historic plantation dwellings in the area by

Figure 2.4 Main house at Friendfield Plantation shown soon after construction, circa 1932. *Source*: Courtesy Athenaeum of Philadelphia, Philadelphia, Pa.

blending the Colonial Revival and Regency styles. A central, hipped roof section features a graceful, curved staircase up to the front door, which is dwarfed by a large stone surround with columns and entablature. A brick pediment with a central hexagonal window and two exterior brick chimneys frame the building. Symmetrical hyphens and flankers pay homage to traditional patterns of massing in the region.[52]

The southern Colonial Revival showed greater attention to landscape design than its northeastern counterpart, whether through misty-eyed descriptions of romantic live oak trees covered in Spanish moss or the examination and creation of formal gardens and former rice fields. Soon after the Civil War, popular literature made images of spacious plantation grounds part of the national iconography. Visitors to the lowcountry found themselves captivated by its landscape, particularly the interplay of built forms and natural features. New owners often hired landscape architects to enhance historic landscape features on their property or to establish a sense of maturity and elegance around new homes.[53] Attention to landscape design developed over time. Initially, few northerners hired professionals to restore or design plantation landscapes. By the 1930s, formal landscape design became an important part of the creation—or re-creation—of showpiece plantations, particularly when plantation houses lacked the "historic" qualities their owners desired.

Loutrel Briggs, a northern landscape architect who designed many gardens for winter residents, has become synonymous with Charleston landscape architecture. During a long and productive career, Briggs designed well over 100 gardens at private residences in Charleston, produced garden and landscaping plans for several shooting plantations on the Cooper River, and designed smaller projects in towns and cities across South Carolina. He carried out only a handful of projects in Georgetown County, none at northern-owned plantations.[54] Instead, the New York firm of Innocenti and Webel served the needs of northerners desiring beautiful landscapes and vistas. Richard Webel and Umberto Innocenti established their firm on Long Island in 1931. Webel tended toward formal, academic designs, while Innocenti approached landscape from a more hands-on, horticultural perspective based on years of working in his family's nursery business in Florence, Italy.[55] Innocenti had an eye for the way plants meshed with the landscape and strong intuitive understandings of relationships created by the middle ground, foreground, and background in his work. Though he did not often formally record his changes, he gained a reputation for straying from the original design to improve upon it.[56] The architects' differences in personality and working style caused few conflicts; their collaboration proved highly successful. As Richard Webel later stated, "my inclination . . . was to extend the architectural design into the landscape [and] Umberto brought the landscape in toward the architecture."[57]

The firm established a set of strong design precepts early on and maintained them throughout its long career. Innocenti and Webel's landscapes are generally axial, often involving long single or multiple rows of trees. Innocenti cultivated relationships with local nurseries, for he and Webel preferred using mature trees in their designs. Their landscapes had an aristocratic, Old World feel and appeared aged and stable upon completion.[58] New plantation owners preferred the firm's formal style and attention to historic details for landscape restoration and the creation of new landscapes that helped new houses fit seamlessly into existing landscapes. Innocenti and Webel's work at Wedgefield and Friendfield plantations offer representative examples.[59]

Wedgefield became one of the firm's earliest projects in South Carolina. William L. Bottomley liked the way Innocenti and Webel's landscapes made his Colonial Revival structures appear more "historic" and frequently collaborated with them on his projects. At Wedgefield, the firm designed a circular drive in front of the house lined with new plantings of mature trees. In addition, they established two more garden spaces: a more open lawn on the east (rear) of the house and a formal horseshoe-shaped garden on the south. Both were framed by short, patterned brick walls.[60]

Although Radcliffe Cheston, Jr., purchased Friendfield Plantation in 1930, he did not think about landscaping until his new house had been completed. In 1936, he hired Innocenti and Webel to restore and beautify the grounds. They created a new driveway lined with magnolia trees that culminated in the carpark before the new entrance to the house. A set of brick gateposts highlighted the entrance to the causeway.[61] Innocenti led restoration of the plantation's unusual sunken serpentine water garden and stabilized a historic brick well. He also identified and highlighted tea plants that remained from eighteenth-century agricultural experiments. To create visual continuity between historical elements and new features, Innocenti oversaw installation of decorative pierced-brick walls with iron gates, brick paths, fish pools, and wrought-iron fountains.[62] Use of such features echoed Loutrel Briggs' designs, which often employed garden walls and walkways built of salvaged bricks to create a ready-made look of age.[63]

The new Colonial Revival-style houses that northerners erected produced mixed results. Academically based reconstruction efforts, such as the Chestons attempted at Friendfield, proved rare. More often, clients simply demanded a Colonial building that seemed appropriate to a setting where a house had previously stood or a new site. Inspiration came from diffuse sources; few owners made historical accuracy a priority.

LOCAL REACTIONS TO THE TRANSFORMATION OF OLD PLANTATIONS

Even sixty years after the Civil War, sectional tensions remained high. In 1912, the nascent Kinloch Gun Club sent a representative to examine several prospective plantations. J. Stuart Groves, a local man, reported to the club president that his representative "did not make a very good impression on the men he met while down here, which seemed to be due to his prejudice against South Carolina."[64] Southerners had little enthusiasm for selling land to outsiders. The attitude of the club's representative only added insult to injury.

Georgetown County newspapers praised "winter colonists" for driving tourism, contributing to the local economy, and saving long-deteriorating houses from almost certain destruction. Generations of sectionalism and the shock of seeing former sites of economic power remade for leisure use fueled tensions. Residents of the lowcountry frequently expressed concern that northerners destroyed game, flouted conservation laws, and wasted valuable agricultural land.[65] Most saw the "invaders" as a mixed blessing. Not surprisingly, locals responded in in multiple ways. Although northerners saved historic buildings and landscapes from loss and subdivision, their actions physically altered the county's heritage and redefined old plantations as recreational venues. The changes they caused proved unsettling to many and angered some.

Mr. Jonathon Daniels, a resident of Charleston, explained the new influx of northern vacationers with this statement: "Rice has vanished, and the boll weevil has slain the cotton. Today . . . millionaires are the successors of rice—they are the cash crop."[66] When relaying the news that Isaac Emerson would be buying Prospect Hill and Oak Hill plantations, the *Georgetown Times* exclaimed, "Good! These rich people only stay here for a few weeks or months during the ducking season, but they spend lots of money. The more the merrier, say we."[67]

Northerners' primary contributions to the local economy lay in seasonal and year-round employment of skilled and unskilled laborers. Each plantation required a year-round superintendent and workers for general maintenance and patrolling the land for poachers. Workers proved all the more important if the owner chose to grow rice or run a small farm, as the Kinloch Gun Club did during its early years. During the hunting season, superintendents hired cooks, laundresses, maids, and hunting guides to ensure the comfort of owners and guests and their success in the field.[68] Workers typically came from families living on plantation lands or nearby. Northerners paid reasonably competitive wages, although better pay at lumber mills and the docks in Georgetown slowly lured some laborers away.[69]

Rankled by their lack of control over the physical past, Georgetown County residents expressed distaste for new social and economic relationships by fighting against the use of formerly valuable agricultural lands as game preserves for the wealthy. Each year, land that had traditionally been shared among neighbors and hunted upon by the same families for generations became restricted as a result of northerners' purchases. In his memoirs, Francis E. Johnstone, Jr., the son of the owner of Belle Isle, candidly stated that he and many of his generation "felt like the ricefields and plantations had been stolen from [their] ancestors." He and his friends responded by poaching and trespassing on their family's former land.[70]

Locals could not help but notice the large numbers of dead ducks and other game that winter residents sent north on refrigerated railcars. Lowcountry newspapers gleefully reported rumors of northerners' overhunting or otherwise refusing to comply with local and state laws. A letter published in the *Georgetown Times* argued that "the greatest menace to our ducks, fur bearing animals, fish and to a great extent our forests, I believe today is the rich men who own these game preserves." This angry citizen contended that soon the fish and game would be all but extinct and criticized wealthy individuals for conspiring with their lawyers to grab land from southerners.[71]

Many felt that seasonal hunting preserves wasted valuable agricultural land. English country estates, as historical precursors to shooting plantations, had traditionally been operated in part as working farms.[72] Old plantations proved attractive to wealthy outsiders because of low land prices, access to wildlife, and strong connections to a supposedly grand agricultural past. Some new owners chose to maintain former outbuildings, grow some rice to attract ducks, and keep livestock, but none intended to make full use of the land's productive capacities. Success in business did not always transfer well to farming. Although many sportsmen and sportswomen came to the lowcountry with plans for working farms, few succeeded. Most soon reduced their efforts to doing only as much as needed to maintain a pastoral effect.[73]

Concern about changing patterns of land use even made its way into national literature. Caroline Lockhart, who later became famous as a Western writer, published "The 'Yankee Snob'" in 1908. The short story featured a former Confederate major who became upset when he found a "Northern Gun Club" had posted "No Trespassing" signs on land adjoining property he owned. When challenged for hunting on the tract by the northern owner, the major stated, "My grandfather and my great-great-grandfather shot over this strip of pine-land, and when I stop shooting over it, it will be because I am too old to hold a gun." By the end of the tale, the two men become friends, a development that gives the narrative a reconciliationist twist.[74] Although fictional, this story would have been intensely familiar to residents of the lowcountry.

The influx of new buyers and new forms of activity increased local interest in the history and current use of old plantations. By the 1920s, writers began crediting northerners with preservation. Perhaps the strongest instances of such commendation appeared in *Carolina Resorts*, a magazine aimed at the wealthy class of people who wintered in the state. The editor argued that without the "peaceful invasion of the Northerners" many fine plantation homes would not have survived.[75] The image of an army of northerners moving through the countryside, restoring treasured houses along the way, almost seemed designed to counter that of Sherman and his armies destroying them.

Of course, recognition of northerners' activities did not always strike an approving tone. The Charleston *News and Courier*, for example, did not refrain from subtly editorializing about the newcomers, even in a series of positive puff pieces about them. Two headlines make their general opinion clear: "Northern Hunters Control Rice Hope" and "Northern Sportsmen Push Inland." The latter article expressed some incredulity that northerners would want to buy land in Williamsburg County, which did not touch the ocean.[76]

Rather than focusing on the question of land ownership, local writers such as Samuel Gaillard Stoney and Alberta Morel Lachicotte sought to claim ownership of regional history. Both generally spoke well of current owners but focused primarily on retelling the histories of local families who had once owned the same land. Stoney also expressed appreciation for the people who "saw in the splendid wreckage of the plantations the background for gracious and comfortable winter homes." He stated that hunters fell in love with irreproducible southern landscapes and then sympathetically restored "many a fine old place that seemed about to vanish forever into the jungle." Stoney's descriptions treated plantations as living beings; like sirens, they lured these men, swayed them into new loyalties to the South, and through their wealth "gained new life."[77]

CONCLUSION

During the past eighty years, progressively fewer of Georgetown County's original plantations have remained intact and in private hands. Some large pieces of land and associated buildings have been preserved or maintained through protective easements, stewardship by private foundations, or the donation of land to the state, but many have been divided into smaller commercial or residential parcels. In the latter case, antebellum history and the enduring romance of Georgetown County's rice plantations and landscapes have been used as a marketing tool.

Whether owned by locals or outsiders, southern plantations continue to draw tourists seeking encounters with an authentic antebellum South. The reasons that remnants of pre-Civil War plantations are present in contemporary Georgetown County—and how the plantation myth has affected modern realities—are ignored or forgotten. Visitors expect historic buildings and sites to act as unbiased storytellers, typically without realizing the full extent of their history. Changes made to aging plantations during the 1910s, 1920s, and 1930s have shaped the way many observers believe southern plantations should look and have affected perceptions of the antebellum past with equal power. Persistence of the romantic mythology that initially played a part in attracting wealthy northerners has kept the history of their activities an open secret in the communities where they lived and played.

When examined individually, the histories of the plantations that northerners turned into grand estates may seem unremarkable or, at the very least, undeserving of close investigation. After all, few parcels of land survive generations of use unchanged, and the same is true for buildings, no matter what their form or use. Yet despite these truisms, on a regional scale, however, the "Second Yankee Invasion" had an astounding influence. Within only a few decades, new owners modified the physical fabric and cultural perception of plantations all along the Carolina coast.

Although the majority of northerners who bought or built plantation houses in South Carolina did not regard themselves as preservationists or revivalists, their architectural impact on the state remains the same. They saved revered houses from ruin and built new plantation-inspired Colonial Revival dwellings that gave old plantations more handsome and stately appearances than had historically existed. Whether winter residents chose to live in old houses or new structures based on traditional styles, they demonstrated a stylistic conservatism that mirrored the architectural preferences of southern communities and popular views of the South and its traditions.

Colonial and antebellum agricultural wealth and the fruits of industrial capitalism are reflected in the majority of the former rice plantations that remain present in Georgetown County. By renovating, restoring, and enlarging old houses, building new mansions, and creating elegant landscapes, northerners gave old plantations new material environments that blended old and new in a seamless fashion. New aesthetics shaped the way people understood the purposes of the new estates and the history of the plantations they replaced. By reshaping the material remains of aging plantations, new owners simultaneously consumed, added new layers to, and refigured the physical history of the South Carolina lowcountry and narratives of its past.

NOTES

1. *Georgetown Times* (Georgetown, S.C.), Dec. 31, 1937, p. 1

2. Samuel Gaillard Stoney, *Plantations of the Carolina Low Country* (Charleston: Carolina Art Association, 1938), p. 42.

3. John F. Ficken, "The President in the South – Fine Sport Yesterday in Shooting English Ducks," *New York Times* (New York, N.Y.), Dec. 20, 1894, p. 1.

4. "Splendid Hunting Along the Santee – Santee Club, Finest in America, has helped to restock the Coast Country," *Georgetown Times*, Oct. 17, 1917, p. 1.

5. A. Radclyffe Dugmore, "A Wilderness Club That Pays its Own Way: The Requirements for a Desirable Club and the Principles of Management for Its Financial Success," *Country Life in America*, Jan. 1906, p. 289.

6. Although this essay focuses on events in Georgetown County, Daniel Vivian's Ph.D. dissertation surveys the development of new estates throughout the South Carolina lowcountry. See Daniel J. Vivian, "The Leisure Plantations of the South Carolina Lowcountry" (Ph.D. diss., Johns Hopkins University, 2011).

7. George C. Rogers, *The History of Georgetown County, South Carolina* (Columbia: University of South Carolina, 1970), pp. 292–293, 487–488. On the decline of commercial rice production, see especially Peter A. Coclanis, *The Shadow of a Dream: Economic Life and Death in the South Carolina Low Country, 1670–1920* (New York: Oxford University Press, 1989), pp. 129–143, 154–156; James H. Tuten, *Lowcountry Time and Tide: The Fall of the South Carolina Rice Kingdom* (Columbia: University of South Carolina Press, 2010), chap. 3.

8. Stuart A. Marks, *Southern Hunting in Black and White: Nature, History, and Rituals in a Carolina Community* (Princeton: Princeton University Press, 1991), p. 48.

9. Frederick Law Olmsted, *A Journey in the Seaboard Slave States; With Remarks on Their Economy* (New York: Dix and Edwards, 1856), pp. 35–37. Olmsted recognized tourism as new in the southern states. He and others criticized the region for underdevelopment. Olmsted saw towns, tourist hotels, and roads as of poor quality. In the early twentieth century, road improvement became a major political issue as civic and business leaders sought to overthrow unfavorable views of South Carolina. Some owners of Georgetown County estates became involved in county and statewide improvement strategies. For example, J. L. Wheeler, a wealthy Pennsylvanian who had made a fortune in silver mining, became a county highway commissioner during the 1930s. Wheeler owned North and South islands, between Winyah Bay and the Santee Delta. See "Dream of Paved Roads Fulfilled," *Georgetown Times*, July 11, 1930, p. 1.

10. Nina Silber, *The Romance of Reunion: Northerners and the South, 1865–1890* (Chapel Hill: University of North Carolina Press, 1993), pp. 67–68, 70–73.

11. "Winter Resorts and Who Attend Them," *The Chautauquan* 16, no. 5 (Feb. 1893): pp. 610–611.

12. Seaboard Air Line, *A Guide to the Famous Hunting and Fishing Grounds of Virginia, North Carolina, South Carolina, and Georgia traversed by the Seaboard Air Line, With a Synopsis of the Game Laws in Those States* (Richmond, Va.: A. Hoen and Co., 1898), p. 7; Southern Railway Company, *Hunting and Fishing in the South* (Washington, D.C.: n.p., 1904), p. 3. For a detailed discussion of the creation of regional character during the nineteenth century, see William R. Taylor, *Cavalier and Yankee: The Old South and American National Character* (New York: George Braziller, 1961).

13. Karen L. Cox, *Dixie's Daughters: the United Daughters of the Confederacy and the Preservation of Confederate Culture* (Gainesville: University Press of Florida, 2003), p. 7; Silber, *Romance of Reunion*, pp. 3–5; David Blight, *Race and Reunion: The Civil War in American Memory* (Cambridge, Mass.: Harvard University Press, 2001), pp. 2–5.

14. Caroline E. Janney, *Burying the Dead But Not the Past: Ladies' Memorial Associations and the Lost Cause* (Chapel Hill: University of North Carolina Press, 2008), pp. 30–37; Cox, *Dixie's Daughters*, pp. 20–27.

15. Silber, *Romance of Reunion*, pp. 39–43.

16. Blight, *Race and Reunion*, pp. 251–254.

17. Coyne Fletcher, "In the Lowlands of South Carolina," *Frank Leslie's Popular Monthly* 31 (1891): pp. 280–288. Historical accounts such as Fletcher's bear some resemblance to the current interest in "ruin porn" or "disaster porn." The proliferation of photo essays and magazine articles featuring dramatic images of Detroit ruins is an example of this phenomenon. See, for example, Mark Binelli, "How Detroit Became the World Capital of Staring at Abandoned Old Buildings," *New York Times*, Nov. 9, 2012, http://www.nytimes.com/2012/11/11/magazine/how-detroit-became-the-world-capital-of-staring-at-abandoned-old-buildings.html?smid=pl-share.

18. James C. Derieux, "The Renaissance of the Plantation," *Country Life*, Jan. 1932, pp. 37–39; Silber, *Romance of Reunion*, p. 69.

19. *Carolina Resorts*, Jan. 8, 1931, p. 4.

20. Michael Bunce, *The Countryside Ideal: Anglo-American Images of Landscape* (London: Routledge, 1994), pp. 2–3, 11, 17, 34–36.

21. Clive Aslet, *The American Country House* (New Haven: Yale University Press, 1990); Richard S. Jackson and Cornelia B. Gilder, *Houses of the Berkshires, 1870–1930* (New York: Acanthus Press, 2006); Daniel DeKalb Miller, *Chateau Country: Du Pont Estates in the Brandywine Valley* (Atglen, Pa.: Schiffer Publishing, 2013); William Morrison, *The Main Line: Country Houses of Philadelphia's Storied Suburb, 1870–1930* (New York: Acanthus Press, 2002).

22. *Georgetown Times*, Jan. 19, 1907, p. 1; "Large Tracts Change Hands – Waccamaw Neck Property is acquired by local lumber dealer," *Georgetown Times*, Feb. 27, 1942, p. 1.

23. "Flames Destroy Baruch Mansion Friday Evening – Occupants Escape Uninjured – Property Loss Totals about $80,000 report – Will Rebuild Soon – Vegetation Practically Unscratched and Outhouses not touched – Baruch family in New York," *Georgetown Times*, Jan. 3, 1930, p. 1.

24. Aslet, *The American Country House*, p. 85; Scott E. Giltner, *Hunting and Fishing in the New South: Black Labor and White Leisure after the Civil War* (Baltimore: Johns Hopkins University Press, 2008), pp. 121–123.

25. William J. Murtagh, *Keeping Time: The History and Theory of Preservation In America*, 3rd ed. (Hoboken, NJ: John Wiley & Sons, 2006), pp. 13–17. For an expanded discussion of the evolution of public and professional practice and theory of restoration, see Jennifer Betsworth, "'Then Came the Peaceful Invasion of the Northerners': The Impact of Outsiders on Plantation Architecture in Georgetown County, South Carolina" (M.A. thesis, University of South Carolina, 2011), chap. 2.

26. Alberta Morel Lachicotte, *Georgetown Rice Plantations* (Columbia: State Commercial Printing Co., 1955), pp. 18–22; *Georgetown Times,* Jan. 2, 1907, p. 1; *Georgetown Times*, Jan. 19, 1907, p. 1.

27. *Georgetown Times*, Jan. 19, 1907, p. 1; Chalmers S. Murray, "Arcadia, Where LaFayette Stopped – House on Waccamaw Peninsula That Ante-dates

Revolution Regains Former Splendor and Takes on Added Beauty and Proportion Under Ownership of Isaac E. Emerson," *News and Courier* (Charleston, S.C.), June 21, 1931, p. B5.

28. Suzanne Cameron Linder and Marta Leslie Thacker, *Historical Atlas of the Rice Plantations of Georgetown County and the Santee River* (Columbia: South Carolina Department of Archives and History for the Historic Ricefields Association, Inc., 2001), p. 78.

29. Lachicotte, *Georgetown Rice Plantations*, 27–29; Chalmers S. Murray, "Ponemah, "Happy Hunting Ground" – Willis E. Fertig, of Pennsylvania, Who Winters on Georgetown Estate, Praises Section for Its Climate, Its Game and Beauty of Black Rivers," *News and Courier*, Apr. 26, 1931, p. A6.

30. John W. McGrain, Brooklandwood (Baltimore County, Md.), National Register of Historic Places nomination, p. 12, http://www.msa.md.gov/megafile/msa/stagsere/se1/se5/003000/003300/003315/pdf/msa_se5_3315.pdf; "Life Visits the Vanderbilt Mansions," *Life* 28, no. 1 (Jan. 2, 1950): pp. 89–92.

31. John F. Harwood, Henry M. Sage Estate (Albany County, N.Y.), National Register of Historic Places nomination, p. 4, State Historic Preservation Office, Albany, N.Y.

32. Lachicotte, *Georgetown Rice Plantations*, pp. 146–147; Chalmers S. Murray, "New Yorker Lives In Coastal Garden – Henry M. Sage Moves Hundred-Year-Old House, Even to Wall Paper, from Newberry to Belle Isle in Georgetown at Cost of many Thousand of Dollars," *News and Courier*, May 24, 1931, p. B6; Floyd Alister Goodwin, *Survival of an Old Rice Plantation: Belle Isle, Georgetown, S.C.* (Baltimore: PublishAmerica, 2009), p. 174. Henry Sage suffered a stroke soon after the beginning of the lease at Belle Isle and died in 1933.

33. Murray, "New Yorker Lives In Coastal Garden."

34. Lachicotte, *Georgetown Rice Plantations*, p. 147.

35. Ibid., pp. 152–154. Henry F. DuPont incorporated large amounts of architectural salvage into his Winterthur mansion. See Jay E. Cantor, *Winterthur: The Foremost Museum of American Furniture and Decorative Arts* (New York: Harry N. Abrams, 1985). Period rooms are a particularly notable, and controversial, example of robbing architectural elements from historic structures. Like the plantation houses profiled here, period rooms have been criticized as imaginative and ahistorical. See Dianne H. Pilgrim, "Inherited from the Past: The American Period Room," *American Art Journal* 10, no. 1 (May 1978): pp. 4–23.

36. Craig A. Gilborn, *Adirondack Camps: Homes Away from Home, 1850–1950* (Syracuse: Syracuse University Press, 2000), pp. 127–128; John Bryan, Gravel Hill Plantation (Hampton County, S.C.), National Register of Historic Places nomination, 2009, http://www.nationalregister.sc.gov/hampton/S10817725011/.

37. "Building Bungalow – Millionaire Sportsman, of New York, Locating winter home at Georgetown," *Georgetown Times,* Dec. 14, 1914, p. 1; Russell M. Doar to Eugene DuPont, Apr. 25, 1917, box 26, Kinloch Gun Club Papers, Hagley Library, Wilmington, Del. (hereafter KGCP).

38. Lachicotte, *Georgetown Rice Plantations*, p. 55; Chalmers S. Murray, "Huntington House Reminds One of Africa," *News and Courier*, June 7, 1931, p. A8.

39. Augusta Owen Patterson, *American Homes of To-Day, their Architectural Style, their Environment, their Characteristics* (New York: MacMillan Co., 1924), p. 59.

40. William Bertolet Rhoads, "The Colonial Revival and American Nationalism," *Journal of the Society of Architectural Historians* 35, no. 4 (Dec. 1976): pp. 239, 242.

41. Kenneth L. Ames, "Introduction" in *The Colonial Revival in America*, ed. Alan Axelrod (New York: W.W. Norton and Co., 1985), p. 10; Mark Alan Hewitt, *The Architect and the American Country House, 1890–1940* (New Haven: Yale University Press, 1990), pp. 83–85.

42. William Bertolet Rhoads, *The Colonial Revival* (New York: Garland Publishing, 1977), pp. 112–114; Hewitt, *The Architect and the American Country House*, pp. 228–230; Catherine Bishir, "Landmarks of Power: Building a Southern Past, 1885–1915," in *Southern Built: American Architecture, Regional Practice* (Charlottesville: University of Virginia Press, 2006), pp. 276–285.

43. Well aware of regional trends, Charleston architect Samuel Lapham created a plan for a contemporary plantation for his Bachelor's thesis. See Samuel Lapham, "A Design for the Family Mansion of a South Carolina Plantation" (B.S. thesis, Massachusetts Institute of Technology, 1916).

44. Southern Farm Agency, "'Hobcaw Barony' – The Finest Game Preserve in the South," *Country Life in America*, Oct. 1904, p. 476.

45. Lachicotte, *Georgetown Rice Plantations*, pp. 12–13.

46. Ibid.; Linder and Thacker, *Historical Atlas of the Rice Plantations of Georgetown County and the Santee River*, pp. 8–50; "Flames Destroy Baruch Mansion Friday Evening – Occupants Escape Uninjured – Property Loss Totals about $80,000 report – Will Rebuild Soon – Vegetation Practically Unscratched and Outhouses not touched – Baruch family in New York," *Georgetown Times*, Jan. 3, 1930, p. 1.

47. John E. Wells and Robert E. Dalton, *The South Carolina Architects, 1885–1935: A Biographical Dictionary* (Richmond: New South Architectural Press, 1992), pp. 94–98; Sarah Fick and John Laurens, Hobcaw Barony (Georgetown County, S.C.), National Register of Historic Places nomination, 1994, p. 10, http://pdfhost.focus.nps.gov/docs/nrhp/text/94001236.PDF.

48. Lachicotte, *Georgetown Rice Plantations*, pp. 13–14; Bernard M. Baruch, *Baruch: My Own Story* (New York: Henry Holt and Co., 1957), pp. 274–275; Chalmers S. Murray, "Baruch Rebuilds at Hobcaw Barony – Financier, Native of South Carolina, Each Winter Comes With Friends of International Note to 23,000-acre Estate in Georgetown," *News and Courier*, May 17, 1931, p. B3.

49. Lachicotte, *Georgetown Rice Plantations*, pp. 137–138; Linder and Thacker, *Historical Atlas of the Rice Plantations of Georgetown County and the Santee River*, p. 546; Frances Cheston Train, *In Those Days: A Carolina Plantation Remembered* (Charleston: History Press, 2008), p. 39; Sandra L. Tatman, "A Study of the Work of Mellor, Meigs, and Howe" (M.A. thesis, University of Oregon, 1977), pp. 59–60; Sarah Fick, Friendfield Nomination (Georgetown County, S.C.), National Register of Historic Places nomination, 1995, pp. 6–7, http://pdfhost.focus.nps.gov/docs/nrhp/text/96000409.PDF; Chalmers S. Murray, "Cheston Builds on Old Plantation – Philadelphian Recreates Atmosphere of Former Historic Francis Withers Mansion on Winter Home at Friendfield on the Sampit," *News and Courier*, Aug. 2, 1931, p. B2.

50. Tatman, "A Study of the Work of Mellor, Meigs, and Howe," pp. 59–60; Train, *In Those Days*, pp. 34, 51; Fick, Friendfield Plantation, National Register of Historic Places nomination, pp. 6–7.

51. Chalmers S. Murray, "Wedgefield Planted for Game – Baltimore Lessee of Black River Lands Cultivates Crops to Attract Deer and Birds – Colonial Dwelling Accommodates Hunters," *News and Courier*, Sept. 27, 1931, p. A9; Linder and Thacker, *Historical Atlas of Rice Plantations of Georgetown County and the Santee River*, p. 444; Lachicotte, *Georgetown Rice Plantations*, pp. 75–77; "To Be the Finest House in Newport – Robert Goelet Will Build a Residence to Cost Millions," *New York Times*, Oct. 18, 1895, p. 6; Susan Hume Frazer, *The Architecture of William Lawrence Bottomley* (New York: Acanthus Press, 2007), pp. 19–20, 28.

52. Frazer, *The Architecture of William Lawrence Bottomley*, pp. 283–287.

53. Francis Pendleton Gaines, *The Southern Plantation: A Study in the Development and Accuracy of a Tradition* (New York: Columbia University, 1924), pp. 13–14.

54. James R. Cothran, *Charleston Gardens and the Landscape Legacy of Loutrel Briggs* (Columbia: University of South Carolina Press, 2010).

55. Gary R. Hilderbrand, *Making a Landscape of Continuity: The Practice of Innocenti & Webel* (Cambridge, Mass.: Harvard University Graduate School of Design, 1997), pp. 12, 18, 23–26, 29–30; Gary R. Hilderbrand, "Richard K. Webel, 1900–2000," in *Shaping the American Landscape*, ed. Charles A. Birnbaum and Stephanie S. Foell (Charlottesville: University of Virginia Press, 2009), p. 365; Gary R. Hilderbrand, "Umberto Innocenti," in *Pioneers of American Landscape Design*, ed. Charles A. Birnbaum and Robin Karson (New York: McGraw Hill, 2000), pp. 192–195.

56. R. Terry Schnadelbach, *Ferrucio Vitale: Landscape Architect of the Country Place Era* (New York: Princeton Architectural Press, 2001), pp. 28, 144; Hilderbrand, "Webel," p. 366.

57. Hilderbrand, *Making a Landscape of Continuity*, p. 16.

58. Hilderbrand, "Webel," p. 366; Hilderbrand, *Making a Landscape of Continuity*, pp. 22, 27.

59. Contemporary project descriptions do not always indicate whether the firm was responsible or just Umberto Innocenti. Innocenti traveled to project sites and, as a result, frequently served as the "face" of the firm. In many cases, primary and secondary sources mention him as the sole landscape architect. For example, Friendfield is spotlighted in a folio advertising Innocenti and Webel, but local primary sources only mention Innocenti. For the purposes of this study, it is assumed that any project Innocenti worked on qualified as a firm project, not a personal job he undertook on his own.

60. Linder and Thacker, *Historical Atlas of Rice Plantations of Georgetown County and the Santee River*, p. 444; Lachicotte, *Georgetown Rice Plantations*, pp. 75–77; Frazer, *The Architecture of William Lawrence Bottomley*, pp. 19–20, 28, 31, 283, 289.

61. Fick, Friendfield Plantation, National Register of Historic Places nomination, pp. 8–9.

62. Train, *In Those Days*, pp. 35–37; Fick, Friendfield Plantation, National Register of Historic Places nomination, pp. 8–9.

63. Cothran, *Charleston Gardens and the Landscape Legacy of Loutrel Briggs*, pp. 79–80.

64. J. Stuart Groves to William Ramsay, Mar. 5, 1912, box 18, KGCP.

65. Similar concerns developed elsewhere. Karl Jacoby's case study of the insider-outsider conflicts over land in the Adirondacks provides an intriguing parallel. See Karl Jacoby, *Crimes Against Nature: Squatters, Poachers, Thieves, and the Hidden History of American Conservation* (Berkeley: University of California Press, 2001), pp. 11–80.

66. William Oliver Stevens, *Charleston, Historic City of Gardens* (New York: Dodd, Mead and Co., 1939), p. 302; "GET SOME YANKEE MONEY," *Georgetown Daily Item*, Feb. 17, 1908, p. 1.

67. *Georgetown Times,* Jan. 2, 1907, p. 1; *Georgetown Times,* Jan. 5, 1907, p. 1; *Georgetown Times,* Jan. 19, 1907, p. 1.

68. J. Stuart Groves to William Ramsay, Nov. 13, 1914, box 20; R. M. Doar to Eugene duPont, Sept. 10, 1917, box 21, both in KGCP; Lee G. Brockington, *Plantation Between the Waters: A Brief History of Hobcaw Barony* (Charleston: History Press, 2006), pp. 54–55.

69. J. Dauforth Bush to William Ramsay, Jan. 20, 1913, box 19; "Report of President on Visit to the Club," Jan. 17, 1916, box 22; both in KGCP.

70. Francis E. Johnstone, Jr., *One Man's Life* (Athens, Ga.: the author, ca. 1993), p. 23; Train, *In Those Days*, p. 51.

71. *Georgetown Times*, June 30, 1933, p. 1.

72. Aslet, *The American Country House*, pp. 14, 141–146.

73. "General Description of the Richfield Property, South Carolina," Mar. 21, 1912, box 17, KGCP. The Kinloch Gun Club expected their property to be self-sustaining but found that it consistently lost money. They retained rice planting and grew produce for onsite consumption but otherwise abandoned commercial agriculture, save for a brief experiment with grapes during Prohibition.

74. Caroline Lockhart, "The 'Yankee Snob,'" *Lippincott's Monthly Magazine* 82, no. 487 (July 1908): pp. 5–79. The universality of conflicts between insiders and outsiders becomes more apparent when Lockhart's background is taken into consideration. Born in Illinois, she lived in Pennsylvania and Kansas for short periods of time before moving to Colorado in 1904. "The 'Yankee Snob'" may have been inspired by local conflicts over land, but she cast the story as a conflict between northerners and southerners to make it more compelling to eastern audiences.

75. *Carolina Resorts*, Dec. 17, 1930, p. 8.

76. Chalmers S. Murray, "Northern Hunters Control Rice Hope – Old Lucas Plantation House in Georgetown Serves as Club for New York Sportsmen Whose Holdings in South Carolina Now Comprise 20,000 Acres," *News and Courier*, June 14, 1931, p. B7; "Northern Sportsmen Push Inland – H.S. Hadden, of New York City, most Recent Addition to Williamsburg Winter Colonists, Old Plantation House Renovated," *News and Courier*, Nov. 22, 1931, p. B14.

77. Samuel Gaillard Stoney, *Charleston: Azaleas and Old Bricks* (Boston: Houghton Mifflin Co., 1939), pp. 24–25; Stoney, *Plantations of the Carolina Low Country*, p. 42.

Chapter 3

Tending the New Old South

Cultivating a Plantation Image in the Georgia Lowcountry

Drew Swanson

In mid-March of 1929, during one of the Georgia coast's most pleasant seasons, a group of newspapermen gathered at Wormsloe Plantation on a press junket. Wymberley W. and Augusta De Renne, the owners of the estate, had invited the reporters to their property to see recent renovations designed to transform a historic sea island cotton plantation into a tourist attraction. The De Rennes led the crowd on a tour of one of Georgia's oldest properties, a piece of land held by the same family since the start of the colony in the 1730s, guiding them through a progression of potential attractions that included a colonial fort, an antebellum house, century-old ornamental plantings, an expansive salt marsh and tidal channels, a slave cabin and cemetery, modern dairy facilities, a library containing a treasure trove of historical documents, and a pair of stately live oak avenues. To seal the deal, the hosts followed the guided tour and note-taking with a buffet luncheon that, as one of the newsmen later wrote, unfolded "in a style redolent of the good old days which won the South her proverbial name for hospitality."[1]

The sales pitch paid off handsomely. A number of the journalists returned to their offices to write articles about the handsome new addition to the state's tourist attractions, repeating Wymberley's and Augusta's descriptions of Wormsloe's charms almost verbatim. Ben Cooper of the *Atlanta Constitution* wrote a representative piece for the state's most important paper. He walked readers around the property, touting the many things to see and the quality people who thought them worth seeing. He noted that the guests who had accompanied the newspapermen included a congressman, two judges, and a county commissioner. According to Cooper the only thing more impressive than Wormsloe's beauty and serenity was the landscape's ability to transport visitors into the mystique of the old planter world. Here, Cooper noted, was a spot where the common person could connect to the grandeur and dignity of the Old South, thanks largely to the generosity of members of that esteemed class. For two centuries Wormsloe had been "preserved against the eyes of the curious" only to be graciously opened "when Mr. De Renne at last yielded to popular sentiment and consented to allow the general public to visit the exquisite gardens of his estate."[2] Gone from this representation of the property was the family's need to make the land profitable in a new way. Equally absent was Augusta. In keeping with the Old South image of the plantation, her role in refashioning the estate lay hidden in Wymberley's paternal shadow.

Other publicity was less stilted but carried much the same message. The local *Savannah Evening News* simply informed readers that Wormsloe was open for business, though it did engage in a bit of long-standing low-country rivalry by declaring that the new attraction might best Magnolia Gardens, an established plantation showplace near Charleston, South Carolina, that attracted a "great horde of visitors" with elaborate gardens and restored buildings.[3] The Wormsloe Gardens that hosted the De Rennes' press junket was one of a growing fold of coastal southern tourist attractions, new

Figure 3.1 An impressive gate opening onto a live oak avenue welcomed visitors to Wormsloe Gardens. *Source*: Francis Benjamin Johnson Collection, Library of Congress, Washington, D.C.

but hardly unique. Like Magnolia Gardens, Wormsloe was a place in transition from the agricultural economy of the nineteenth-century lowcountry to the service and tourism economy of the twentieth, a shift that challenged but ultimately reinforced traditional landscape uses and the self-identities of white landowners, a group long accustomed to conflating planting and power. The founding of Wormsloe Gardens, then, was part of a larger remaking of the South Carolina and Georgia lowcountry from a land of rice, sea island cotton, slaves, and shipping to a coastline of recreation and historical tourism.

Two forces at work in the emerging tourist South of the interwar years proved especially influential at Wormsloe. The first was the ongoing effort to redefine the meaning of the term "plantation." Before the Civil War, plantations had been relatively straightforward; they combined agricultural production with unfree labor. And that agricultural production usually meant commodity crops for distant markets. Additional details might be debated endlessly, but they mattered little in comparison to the labor-production union. Perhaps emancipation meant that planters became landlords rather than labor lords, but this maxim obscures as much as it illuminates.[4] Although useful for the economic truths it conveys, the argument is culturally deceptive. Antebellum planters often cared intensely about their land—Wormsloe's masters certainly thought themselves landlords before and after the war—and southern elites quickly devised methods of securing labor after emancipation that afforded nearly as much control over workers as chattel slavery had before. Apprenticeships, lack of economic opportunity, convict leasing, and various forms of debt peonage all proved brutally effective at yoking black labor to rural lands owned by whites.[5]

For a time, at least, production for markets proved the more challenging part of the equation. Overall staple production in the South dipped immediately following the war but then grew in the late nineteenth and early twentieth centuries and eventually far outstripped the most productive antebellum years.[6] Centers of staple production shifted, producing a new geography of southern agriculture that left some former agricultural hubs behind, especially along the southern Atlantic coast, where rice became less feasible without slave labor (thanks in part to the increasing demand for black labor in more prosperous areas, such as the Mississippi Delta's alluvial cotton lands). How would the old masters of the rice coast maintain their plantations—and, by extension, their status as planters? As the ground shifted in the decades flanking the turn of the century, understandings of people and place hung in the balance. What could a plantation be, once agriculture ceased to be definitive? Wormsloe's masters took up these questions and found answers in history. They accepted that change must come and worked to create a landscape that seemed as much a part of the past as the present, where the labor of production was obscured (and thus issues of race hidden), and memory focused on the fruits rather than the mechanics of mastery. In this New South, a plantation became a place where a planter lived, no matter his or her current source of income.

The influence of northern capital and ideas also proved crucial. Northerners imagined plentiful opportunities in the post-Civil War plantation South. Some sought direct economic benefit by assuming the role of planters themselves. Others simply came as tourists to see and experience an exotic, less "American" place.[7] Northern money and people influenced plantation landscapes of all sorts. Wormsloe Gardens offers one example. As the De Rennes struggled to continue what they saw as a plantation tradition in the face of changes wrought by emancipation and shifting patterns of agricultural production, they also recognized that northern fascination with southern plantations presented opportunities. In places such as the lowcountry, southerners' reimagining of plantations depended a great deal on northern interest, a fact that Wymberley and Augusta recognized as they worked to turn Wormsloe into a showplace.

Wormsloe's saga may hardly have been unique, but its exceptionally rich history provided ample fodder for Wymberley and Augusta's efforts. Wormsloe represented a veritable showcase of Georgia's past. Wymberley's great-great-great grandfather, Noble Jones, had arrived with James Oglethorpe and Georgia's first English colonists in 1733. A man of some standing, Jones served as a surgeon and surveyor for the infant settlement. Jones was also a military man, charged with constructing and garrisoning a fort southeast of Savannah that could guard one water approach to the city against Spanish incursion. It was this fort that brought Jones to Wormsloe. The colonial trustees granted the plantation on the Isle of Hope to Jones for use as a fort site and a private estate. Jones grew wealthy at his many posts, acquiring dozens of slaves once the practice became legal and amassing additional acreage in the lowcountry and the Piedmont. Noble's son, Noble Wimberly, also became an influential Georgian. Politically active, he served in the provincial legislature at the outbreak of the American Revolution and participated in the convention that drafted the state constitution, a role that earned him the sobriquet "Morning Star of Liberty." After the war, he remained active in Savannah city politics and became a prominent physician.

George Jones followed in his father's footsteps by becoming a respected doctor and assuming an active role in several local organizations. His son, George W. Jones (who changed his surname to De Renne following the Civil War), became a noted agricultural reformer, planter, and prominent collector of historical books and documents. Wymberley Jones De Renne, the next generation of the family to live at Wormsloe and Wymberley W.'s father, was also a well-traveled historical collector and a writer, bibliographer, and farm manager. Wymberley J. continued the family's long legacy of supporting

Savannah civic and social clubs by helping to steer the activities of organizations such as the Georgia Historical Society and the Chatham Hunt Club.

Each generation of the Joneses-turned-De Rennes had believed Wormsloe Plantation a special place, a sort of feudal seat where gentleman patriarchs presided over a growing empire of land, wealth, and influence. For Noble, Wormsloe served as his first plantation and part of his legacy; for Noble W. it was a country seat and an escape from the political realm; for George W. it became a home place and the center of his agricultural endeavors, a plantation proper as people of the era understood the term; and for Wymberley J. it was both home and patrimony. For almost two hundred years, then, Wormsloe's masters had layered meaning after meaning upon the landscape, which in turn led each of them to reflect upon and add to those they inherited.[8]

By the time the newspapermen tramped Wormsloe Gardens and the grounds opened to tourists with regular hours in March of 1929, the De Rennes had created a standardized visitor experience, a looping path through the estate that began and ended at massive concrete and iron gates at the head of a live oak alley more than a mile long. Entering the main drive, guests passed "through the museum of wild birds" in the gatehouse (a taxidermy collection assembled by Wymberley's father), then moved south along the oak alley to the tabby ruins of Noble Jones's original fort. Next they strolled past an old mulberry tree that reputedly dated to the colonial era, when law required each settlement to plant mulberries as feed for the silkworms the trustees believed would make the colony profitable. Moving from the colonial lowcountry to the antebellum cotton kingdom, the next stop on the loop was a restored slave cabin from the 1850s, complete with reproduction slave furniture and tools "made by one of the darkies of the old school who has learned his trade from methods handed down from the original carpenters," drying vegetables hanging from the walls, and a local woman playing the part of mammy, tending the fire and dispensing plantation lore (in short, a living history exhibit conforming to the standards of the time). Next on the itinerary was the slave cemetery, but here the garden interpreters practiced more restraint: "The only vestiges of the original burying grounds are the mounds which rise indefinitely here and there, occasionally marked by a simple wooden marker or the relics of old pottery." A detour took curious visitors to see earthworks that had sheltered a Confederate artillery battery during the Civil War, followed by a walk through banks of colorful azaleas. The tour culminated with formal walled gardens (designed by Augusta) just south of the house, a glimpse of a stand-alone library that housed the family's collection of historical documents, and the antebellum mansion itself. Before making their way out the drive and ending the day, tourists had opportunities to purchase tinted postcards depicting Wormsloe scenes and to enjoy refreshments at a tea room.[9]

Hesitant to rely on the reporters alone, the couple advertised the new garden in local, state, and regional publications. Touting Wormsloe's history and its horticultural attractions, they proclaimed it a site of unique "Old world charm and historic lore in the South's most, [*sic*] naturally beautiful estate."[10] Visitors could immerse themselves in landscapes evoking the colonial era, when the fragile Georgia colony feared Spanish invasion, or they might relive the Civil War naval blockade, all while admiring gardens with "thousands of azaleas and hundreds of camellias besides a wide collection of native shrubs, evergreens and trees."[11] Wormsloe quickly became one of Savannah's most popular tourist sites. By 1935 the plantation was a staple on the city's garden and homes tours.[12]

The Wormsloe Gardens that took shape in the 1920s largely resulted from Augusta's vision. She had long followed horticulture and design. When she began laying out formal gardens on the Wormsloe grounds in 1919 (within a few years of Wymberley J. De Renne's death and her and Wymberley W.'s move to the plantation), she built on the older work of her father-in-law.[13] Eventually she would come to be so closely associated with her Wormsloe gardens that her obituary would privilege her horticultural efforts above all of her other accomplishments.[14] Wymberley also had an interest in gardening, but it was no accident that the plantation's horticulture fell under Augusta's sphere, for she possessed an intense passion for the enterprise. Acting as a precocious horticultural hobbyist, Augusta entered the realm of professional (ostensibly male) science and engaged in garden-related correspondence with botanists at the Bureau of Plant Industry, ornithologists (inquiries about establishing a bird sanctuary), nurserymen, and fellow garden designers across the country.[15]

Augusta's public life was not singular; for all the overt masculinity surrounding the development of professional disciplines at the turn of the twentieth century, the manipulation of nature through horticulture, landscape architecture, and gardening offered women opportunities for influence, perhaps because of a growing Progressive association between well-ordered landscapes and well-managed homes.[16] At times, characterizations of horticultural work as women's labor, soft and feminized in a way that fit poorly within a new world of rigid disciplines, carried unfortunate connotations for the burgeoning profession.[17] Yet women throughout the nation leveraged their garden work into broader forms of civic engagement, partly because horticultural activity became part of other club work and partly because garden clubs often morphed into pseudo-political bodies. Some women, such as Beatrix Jones Farrand, carved out space for themselves as certified horticultural experts or landscape architects, complete with handsome clienteles.[18] Augusta also crossed these fuzzy barriers. Her work as a semiprofessional garden designer overlapped with other civic engagements, which variously included memberships in the Junior League and the Garden Club of America and service as the first president of the Savannah League of Women Voters and as the first vice president of the Garden Club of Georgia.[19]

Central to Augusta's garden design was the incorporation of historical arti-facts. She believed Wormsloe Gardens should not only convey history but be made from it. To this end, she diligently collected historic materials and used them to frame the garden's architectural bones. Enclosures "made of bricks from the old slave quarters" divided the formal walled garden.[20] She paved the walkways that led guests past flowerbeds with ballast stones brought to Georgia by colonial English ships. Sunlight pierced the higher brick walls through openings that framed salvaged iron work from dismantled lowcountry plantation homes. A vanished Sapelo Island estate furnished marble columns that became focal points set amid formal plantings. And a wellhead from yet another defunct plantation created another object of interest.[21] Farther afield, the couple oversaw the remodeling of the plantation's sole surviving slave cabin, which they surrounded with banks of azaleas and camellias.[22]

A combination of tradition and necessity inspired the formalization of the gardens. Curious visitors had long toured the De Rennes' home and its grounds. During the nineteenth century, Wormsloe's masters had proudly shown off the estate and hosted local civic functions where they extended the expected planter hospitality to guests. During the early twentieth century these activities expanded. Wymberley J. De Renne's historical collections attracted interest and Wormsloe's horticultural reputation grew. By 1924, Wymberley W. and Augusta employed a formal register to keep track of plantation guests, some of whom came from as far afield as New York and England. In 1925 they began annually opening the grounds to paying visitors, with the proceeds benefitting local charities.[23]

Economic misfortune transformed the nascent tradition of annual openings by the end of the 1920s. Wymberley and Augusta found themselves deep in debt and worried about the future of Wormsloe. In a 1929 memorandum, Wymberley noted that his outstanding debts totaled almost $530,000 (equiva-lent to more than seven million 2013 dollars).[24] Real estate investments in Savannah had not proven profitable, and the impending economic collapse would make reversing the financial tide even more difficult. Concerned for her brother and for the future of the family estate, Wymberley's sister, Elfrida, secured some of the debt against the plantation by purchasing a mortgage, a measure that allowed Wymberley and Augusta to remain on the property. The couple pursued various options for reducing their debts, including renting Wormsloe's agricultural land to the Foremost Dairy Company, which sold milk, cream, butter, and cheese to Savannah customers. By 1934 the situation had deteriorated further, prompting Elfrida to seek other ways of protecting the plantation. Her most ambitious scheme sought to rent Wormsloe to a

New Yorker who offered to pay $10,000 annually to live on the estate and to spend an incredible $60,000 in upkeep and improvements. Elfrida pushed her brother to move out, promising in exchange to forgive the mortgage at the end of the lease and to deed the plantation to Wymberley's children, but he refused. Both siblings seem to have genuinely believed themselves to be acting in the best interests of the family legacy and desired to keep Noble Jones's land in the hands of his descendants. Turning Wormsloe Gardens into a tourist attraction, then, owed a great deal to Wymberley and Augusta's desperate circumstances and their pursuit of financial stability.[25]

These particular financial tensions may have been new, and unique to the situation, but the larger tension of southern adjustment to the economic realities of twentieth century America framed this smaller drama. Wormsloe had rarely been very profitable. Even in the cotton boom days of the 1850s, the De Rennes realized losses rather than profits in some years, and following emancipation Wormsloe's masters had struggled through a variety of schemes to make the land pay without slave labor. George W. De Renne and his son alternately rented portions of the estate to other white landowners and to former slaves and tried various forms of agriculture ranging from dairying to truck farming. Increasingly Wormsloe became a place of recreation and leisure more than a site of profit. By the turn of the century it was a landscape that cultivated the planter image more successfully than plantation crops.[26] Wormsloe's struggles symbolized the shifting agroeconomy of the postwar South, as some regions—the newly reclaimed wetlands of the Mississippi and Arkansas Delta, the Texas cotton belt, and the sandy tobacco lands of the eastern Carolinas, for example—became more productive and profitable, often at the expense of older plantation regions, the lowcountry among them.[27] Although few descendants of lowcountry planters found themselves in circumstances exactly like Wymberley and Augusta's, many landowners faced the basic problem of what to do with agricultural landscapes that no longer turned a profit under traditional uses.

As they reshaped Wormsloe's grounds and crafted their interpretation of the plantation's history, the couple tapped into long-standing traditions of memory-making and landscape-building on Wymberley's side of the family. What seemed new was in important ways a reflection of previous efforts. Wormsloe as a plantation showpiece and a repository of state history had a well-established legacy. Family members had long been bibliophiles of some standing. George W. Jones amassed an impressive collection of colonial books and documents before the Civil War, only to have it burned by the Union cavalry that briefly occupied the plantation in 1864 during General William T. Sherman's march to the sea. George's son, Wymberley J., rebuilt the collection and even expanded it, adding such important pieces as an original copy of the Confederate Constitution. The denouement of his collecting

Figure 3.2 Surrounded by ivy and Spanish moss, the De Rennes' historic plantation house stood at the center of Wormsloe Gardens. *Source:* Francis Benjamin Johnson Collection, Library of Congress, Washington, D.C.

came in 1909 with construction of a dedicated freestanding library on the plantation grounds. Modern and "fireproof," the brick and concrete pile was an impressive structure that looked out over the salt marsh and housed almost two centuries of state and southern history.[28]

At times the lure of the family's collection extended its reach, entangling the history of the plantation and the interests of the Joneses/De Rennes with national historiographical impulses. One such moment occurred in the spring of 1909. A young University of Wisconsin historian, Ulrich Bonnell Phillips, had been involved with the Georgia Historical Society in an effort to assemble a comprehensive bibliography of the state's history. The project drew him into the same orbit as Wymberley J. De Renne, whose collection had grown to be the most comprehensive on the subject. In 1909 Phillips visited Savannah, primarily to spend time exploring the treasures held in Wormsloe's new library.[29] While examining the De Renne collection, Phillips gave a public lecture in town. His talk encapsulated an early version of the history of slavery that he had started piecing together from archives across the United States. A reporter from the *Savannah Morning News* summarized Phillips's message in an article title, labeling slavery an "Influence for Good of [the] Negro," and elaborating in the lines that followed, "Master Was Compelled by His Position to Be Considerate, Standing Between Tyranny and License." Slaveowners, the article noted, "Did Much Same Work That Is Now Being Done by Universities in Social Settlement Work in Cities."[30] Phillips's speech presented the argument of his magnum opus, *American Negro Slavery*, which appeared nine years later and, in combination with William Dunning's work on Reconstruction, anchored historians' apologies for slavery during an era of growing reconciliation between North and South.[31]

A few years later, another historian of national import linked his work even more directly to Wormsloe and its influence on the historical profession. Ellis Merton Coulter, a professor at the University of Georgia and author of dozens of books on the South, began a correspondence with Elfrida Barrow, Wymberley's sister, in the mid-1920s. He subsequently worked extensively in the De Renne library. Twenty years later, Coulter formalized his connection to Wormsloe and its historical legacy by beginning a book on the first three generations of the Jones family in America. He published the study as *Wormsloe* in 1955.[32]

The Jones/De Renne family's legacy of horticultural pursuits grew to be as rich as its historical preservation efforts. Each generation had taken pleasure in experimenting with lowcountry soils and seasons in trials of an astonishing range of plants. In addition to his other positions, Noble Jones served as the Georgia colony's first official forester and planted mulberry trees for sericulture, citrus fruits, and other subtropical species in his efforts to understand and master the local climate. His son, Noble Wimberley, also possessed a great botanical curiosity, corresponding with such luminaries as Benjamin Franklin, who sent upland cotton and Chinese tallow tree seeds to Wormsloe. During the antebellum era, George W. Jones embraced the techniques of "scientific" agriculture. He read agricultural journals, ran trials on a wide range of crop varieties, and kept careful records of his work. He also beautified the Wormsloe grounds with ornamental plantings and planned gardens. Wymberley Jones De Renne continued and expanded upon his father's work, adding thousands of ornamental plants and new species to the estate grounds. Some of the specimens he added came from an elaborate coal-heated glass greenhouse that he had built. The culmination of this grand planting regime was the construction of the plantation's impressive live oak avenue. More than a mile long and arrow-straight, it sought to invoke the grandeur of the Old South. Views of the avenue have since become the iconic image of the site. Under the family's care, camellias, azaleas, palms, roses, and hundreds of other flowering, showy, or otherwise interesting specimens annexed more and more of the historic cotton plantation, turning a landscape once dominated by the white "bloom" of the cotton boll into one where kaleidoscopic vistas of blossoms reigned supreme.[33]

During the 1930s, Wymberley and Augusta used the gardens to shape their own version of the plantation's stories and scenes. Their efforts echoed broader impulses. The Great Depression turned society upside down, with ripples that were as much psychological as economic. The most famous expressions of the Depression's effect on American thinking were the futurist planning programs of Franklin Delano Roosevelt's New Deal, such as the Resettlement Administration's attempts to categorize the best use of particular lands and move families around the country, the conservation projects carried out by the Civilian Conservation Corps, and the Tennessee Valley Authority's mission to "modernize" a region roughly the size of the United Kingdom through rural electrification and allied work.[34] The Depression proved equally powerful in fostering contemplative reflection and nostalgia, however. Many Americans found themselves longing for the "good old days" when the economy had been less volatile and life simpler (usually without considering whether such days had ever existed). Here the Works Progress Administration (WPA) often took the lead. In addition to employing skilled but out-of-work Americans on a wide range of public works projects, the WPA also found work for unemployed writers, artists, and musicians. The agency's cultural work often focused on local history, especially collecting and compiling various works of history and folklore. To this end WPA workers interviewed elderly ex-slaves, recorded Appalachian folklore and music, catalogued historical records in dusty rural courthouses, and wrote guides to historical sites.[35]

The Georgia lowcountry drew a great deal of attention from WPA historians and writers, thanks in part to its strong regional identity and plantation history and, moreover, a high rate of unemployment. Under the supervision of Mary Granger, the Georgia Writer's Project collected oral histories and folklore from African Americans living in Savannah and surrounding districts. Focusing on the accounts of ex-slaves and particularly on religious practice and superstition, Granger's folklorists wrote of a mystical and perhaps mythical nineteenth-century lowcountry centered on the rhythms of plantation life and surviving transatlantic cultural linkages.[36] Project writers also produced overtly commercial work such as *Savannah*, a guide to the city's historic attractions that contained tour itineraries for a number of outlying plantations, Wormsloe included.[37] The Savannah unit of the Georgia Writer's Project even began a history of Wormsloe, compiling nearly fifty pages of historical notes that built upon Wymberley and Augusta's work. Its research helped Coulter complete his study of the estate and the Jones family almost twenty years later.[38]

The interwar period also made historical and garden tourism more practicable and profitable because of transportation improvements. The lowcountry benefitted from an influx of wealthy northeastern tourists who rode coastal rail lines south during the winter months or traveled by ship. Middle-class travel via growing rates of automobile ownership proved more important still. The construction of the Dixie Highway, a network of modern paved roads that connected the industrial cities of Ohio, Michigan, Illinois, and Indiana to Florida, had direct benefits for Wormsloe. One branch brought auto tourists through Savannah.[39] Although family traditions and financial necessity supplied the most important impulses behind the establishment of Wormsloe Gardens, technological advances such as the Dixie Highway added to its success.

Other lowcountry developments made the promise of tourism apparent as well. Georgia's sea islands in particular proved popular as sites of recreational development, particularly for elite northern businessmen and their families and friends. A list of these "snowbirds" quickly grows long, but a few examples illustrate the extensive investments that outsiders made in the region by the Wormsloe Gardens era. Henry Ford bought nearly 70,000 acres in the lowcountry during the 1920s and 1930s; the Wanamakers of Philadelphia transformed Ossabaw Island into a hunting resort; R. J. Reynolds Tobacco Company scion Richard J. Reynolds, Jr., retreated to Sapelo Island; a combination of northern and southern investment turned Jekyll Island into a prestigious social club catering to the fabulously wealthy; and Pittsburgh's Carnegie family owned a share of Cumberland Island. A similar situation prevailed along the South Carolina coast. Looking north and south, the De Rennes could not help but see tourism as an increasingly attractive opportunity.[40]

The format that the couple settled on—a gracious plantation home surrounded by elaborate gardens and framed in memories and myths of the Old South (and a few actual pieces of that South in the form of ironwork and salvaged columns)—was an all but standardized "type." In addition to the Savannah-area sites highlighted by WPA writers, the aforementioned Magnolia Gardens, located outside Charleston in the South Carolina low-country, offered an example of what Wormsloe might become. Magnolia and Wormsloe had remarkably similar pasts. The Drayton family established Magnolia in the early years of the Carolina colony and had held it through succeeding generations for almost two hundred and sixty years by 1930. It had been a rice plantation worked by slave labor prior to emancipation; after the Civil War, the Draytons increasingly pursued ornamental horticulture in an effort to make Magnolia a site of recreation and leisure. They opened the estate to paying tourists in 1872 and, during the early twentieth century, its popularity grew thanks in part to the same developments that benefited Wormsloe.[41] Despite the "extravagant price of admission," Magnolia drew rave reviews. Edward Twig, a writer who gained notoriety for attacking the "myth" of Charleston gentility in one essay, held a soft spot for Magnolia, which he labeled a landscape "more ethereal than any man has yet been able to express."[42] The De Rennes thus had firm models to draw from as they shaped Wormsloe in the mold of 1930s southern historical tourist culture.

The De Rennes' efforts did not achieve success easily. Much like the labyrinthian tidal waterways surrounding Wormsloe, the stream of memory making followed more than one channel at any given time. For all of Augusta and Wymberley's success shaping historical interpretation and reshaping the plantation landscape to suit those interpretations, alternate histories took shape simultaneously. Just beyond Wormsloe's northern boundary, Savannah's suburban sprawl lapped against the lore of colonial and ante-bellum history in the form of Dupon subdivision, part of the growing Isle of Hope village, a community of globally inspired architecture, yachting aficionados, and city workers who commuted to Savannah on the City Suburban Railway Line.[43] At the estate itself, the De Rennes relied on African American workers to execute their plans and portrayals, from a costumed mammy who served refreshments to visitors to guides who led parties across the grounds while describing the various plantings and historical sites. Other black workers maintained Wormsloe by trimming shrubs, pulling weeds, painting, and taking care of the owners themselves. African American labor at once helped portray Wormsloe as a vestige of the past and challenged the idea of omnipotent white expertise by demonstrating the skills and abilities of black minds and hands.[44] Moreover, while the estate grounds may have struck some visitors as infinitely malleable, they proved less than infinitely tractable. Tropical storms twisted and uprooted trees, insects attacked plantings, and the tidal creeks and marshes surrounding the plantation continued their eternal shifting, literally expanding and eroding the margins of the family's physical legacy. Other humans exploited the remade landscape for their own uses. Illicit distillers, for example, set up operations in the thickening woodland

of the plantation's margins, as forest re-covered old cotton fields. In 1935
bootleggers even stole one of the De Rennes' automobiles in an unsuccessful
effort to move a large still.[45]

Despite challenges to the De Rennes' agency, their efforts to remake the plantation into a tourist grounds were—at least temporarily—successful, and they
can hardly be construed as unique. But why did Wormsloe Gardens matter?
After all, it proved relatively short-lived. The income that Wymberley and
Augusta drew from admission fees, postcard sales, and tea room concessions
did not offset their financial liabilities, even in combination with their other
activities. By the late 1930s Elfrida Barrow felt forced to take possession of
the plantation to prevent its sale outside the family, and the De Rennes moved
to Athens, Georgia, though they always retained strong feelings of affection
for Wormsloe and the gardens they built. Seeking greater privacy Elfrida and
her husband Craig closed the grounds to public admission. But the Wormsloe
Gardens period did leave a lasting stamp on a landscape that shaped regional
and state historical memory in the years to come. The Barrows continued to
host visitors in significant numbers in the following decades, reverting to the
pre-Gardens practice of showing off family history and the grounds to invited
guests and using Wormsloe for a wide variety of civic programs.

Wormsloe's period of greatest significance for lowcountry historical interpretation began in the 1970s, when the historic plantation became a state
historic site charged with preserving Georgia's colonial heritage. During the
1950s and 1960s the Barrow family faced an escalating property tax burden
and legal wrangling with the Chatham county government over tax rates
that eventually reached the state supreme court. Fearing that the old estate
might be broken up, the Barrows approached state authorities about a possible sale. For its part, the state enthusiastically favored making Wormsloe a
public historical park. With Governor Jimmy Carter lobbying for acquisition
and the Nature Conservancy brokering the transfer, the state took control of
most of the historic plantation on the last day of 1972 (the family reserved
the plantation house and a small tract of land for personal use). Wymberley
and Augusta's historic preservation and horticultural work became integral
parts of the Georgia Department of Natural Resources' management of the
property.[46]

On a larger stage, Wormsloe Gardens was an exemplar of a wave of historic landscape management efforts in the early to mid-twentieth century,
as the nation rushed to modernity, sought to cauterize the wounds of the
Civil War once and for all, and increasingly placed historical interpretation
in the hands of professionals. Augusta and Wymberley's work found parallels across the South (and beyond), not just on other lowcountry estates like
Magnolia. A few examples suffice. Women's groups in Natchez, Mississippi,

put on historical pageants, restored cotton grandees' homes and gardens, and established their city as the belle of the old cotton South.[47] In the North Carolina Piedmont, Katharine Reynolds, wife and business partner of tobacco tycoon R. J. Reynolds, turned an old tobacco farm into Reynolda Estate, a landscape where she worked much as did Augusta to build a showplace that combined horticulture with modern scientific agriculture and celebrated local history and society while attempting to instill virtue in others. Reynolds invited tour groups, consulted scientific experts, and delved into landscape architecture.[48] On the edge of the Georgia Mountains, interested history buffs and state authorities worked to restore the Chief Vann House and grounds, once the home of James Vann, a prominent Cherokee chief and planter, as a place where visitors could at once engage in "playing southern, [and] playing indian."[49] In New Orleans, civic boosters worked diligently to erase perceptions of the port as a den of vice and instead fashion a new image heavily reliant on portrayals of a storied southern past.[50] At Williamsburg, Virginia, John D. Rockefeller, Jr., poured millions of dollars into reconstructing the colonial capitol, creating a townscape that was at least as much fiction as fact.[51] And in southern Pennsylvania in 1933, the National Park Service (NPS) acquired the nation's most famous battlefield, Gettysburg, and set to work creating a pastoral landscape that sought to use nature as a vehicle for time travel and convey certain interpretations of the Civil War's seminal battle.[52]

In all these instances, memory and landscape making was as much a process of erasure as construction. No example is more telling than the development of the national parks following passage of the 1916 Organic Act that created the National Park Service. Out west the new NPS labored to remove all traces of Indian presence from ostensible wilderness, blocking historic Blackfeet rights within Glacier National Park and trying to figure out what to do with the Sierra-Miwock people living in the Yosemite Valley.[53] In the east, in both Great Smoky Mountains National Park and Shenandoah National Park, the NPS simultaneously sought to cultivate idealized wild landscapes while shaping historical interpretation. In both instances the idea of primitive, rustic Appalachia—a land of log cabins, Elizabethan English, and contemporary ancestors—proved irresistible. The NPS tore down modern frame houses, restored surviving log structures, dismantled electrical grids, moved roads, restored water-powered gristmills, and created suitably "rustic" parkscapes. In Glacier, Yosemite, Shenandoah, and the Smokies, as at Wormsloe, kernels of historical truth became the cornerstones of historical interpretation and landscape management while other pasts were obscured. In the end, a combination of reshaping and retelling laid the foundation for the future of these landscapes.[54]

At Wormsloe and at similar history/nature/tourism sites across the South, the past was written onto the landscape, only to be subsequently reshaped

Figure 3.3 Among Wormsloe Garden's scenic attractions was a colonial fort that had once guarded Savannah from potential Spanish attack. *Source*: Francis Benjamin Johnson Collection, Library of Congress, Washington, D.C.

and then carried forward. Histories were as much made as preserved. The temptation is to regard these places and their historical interpretation as "fake," something less than authentic, crafted to deceive.[55] Such a conception is ultimately unsatisfying, however. For all their fabrication, these landscapes of memory were the work of real people, people like Katharine Reynolds, members of the New Orleans Chamber of Commerce, John D. Rockefeller, Jr., and Augusta De Renne. Over time, such places have acquired a virtual "agency" of their own, conveying meaning, symbolism, values—history—to those who engage in historical tourism. To cast these sites aside as "fake" is to ignore their real power, and thus to ignore an import force in forming understandings of the past.

Wormsloe Gardens and similar early twentieth-century southern sites might be more productively imagined as places of intersection, loci of various strains of history merging to create something new. On the old lowcountry landscape two pasts met: one was the actual history of a cotton plantation struggling with the economic and social realities of a post-emancipation world, the other was post-Civil War southern identity making, busily building a moonlight and magnolias memory of the antebellum South. The resulting fusion, which seamlessly melded Wormsloe's history and landscapes into the tales of Phillips and Coulter, began building its own history almost immediately. Visitors absorbed this new past and dispersed it as they left the plantation grounds. The resulting history reflected the desires of the tourist gaze as surely as the boxwood hedges and banks of azaleas conveyed the tastes and dreams of Wormsloe's old masters.

NOTES

1. "Newspaper Men See Wormsloe in All Its Beauty," *Savannah Press* (Savannah, Ga.), Mar. 14, 1929 (clipping), George Wymberley Jones De Renne Family Papers (hereafter GWJD), box 31, fol. 3, Hargrett Rare Books and Manuscript Library, University of Georgia, Athens, Ga. (hereafter Hargrett).

2. Ben Cooper, "Glorious Gardens, on Stately South Georgia Estate Are Opened to Public after 200 Years of Privacy," *Atlanta Constitution* (Atlanta, Ga.), Mar. 24, 1929 (clipping), fol. 3, box 31, GWJD.

3. "Wormsloe and Magnolia Gardens," *Savannah Evening News* (Savannah, Ga.), Mar. 22, 1929 (clipping), fol. 3, box 31, GWJD.

4. The framing language here is James L. Roark's, in "From Lords to Landlords," *Wilson Quarterly* 2, no. 2 (spring 1978): pp. 124–134.

5. The body of literature on postwar white southerners' efforts to re-secure black labor is vast. For selected firsthand accounts demonstrating early forms of postwar plantation labor, see Steven Hahn et al., eds., *Freedom: A Documentary History of Emancipation, 1861–1867*, Series 3, Volume 1, *Land and Labor, 1865* (Chapel Hill: University of North Carolina Press, 2008).

6. In 1860 the U.S. produced 434,209,462 pounds of tobacco, 5,387,052 bales of cotton, and 187,167,032 pounds of rice, most of it in slave states. By 1920 the figures had risen to 1,372,993,261 pounds of tobacco, 11,376,130 bales of cotton,

and 35,330,912 bushels of rice, most still raised in former slave states (California was not yet a major producer of rice or cotton in 1920). See Joseph C. G. Kennedy, *Agriculture of the United States in 1860* (Washington, D.C.: Government Printing Office, 1864), pp. xciv, xcvi, 185; *Fourteenth Census of the United States Taken in the Year: 1920*, Vol. 5, *Agriculture* (Washington, D.C.: Government Printing Office, 1922), pp. 771, 832, 838.

7. Good places to start on these two expressions are Lawrence N. Powell, *New Masters: Northern Planters During the Civil War and Reconstruction* (New Haven: Yale University Press, 1980); Nina Silber, *The Romance of Reunion: Northerners and the South, 1865–1900* (Chapel Hill: University of North Carolina Press, 1993).

8. For detailed histories of Wormsloe and its masters prior to the 1920s, see Drew Swanson, *Remaking Wormsloe Plantation: The Environmental History of a Low-country Landscape* (Athens: University of Georgia Press, 2012); William M. Kelso, *Captain Jones's Wormslow: A Historical, Archeological, and Architectural Study of an Eighteenth-Century Plantation Site near Savannah, Georgia* (Athens: University of Georgia Press, 1979); William Harris Bragg, *De Renne: Three Generations of a Georgia Family* (Athens: University of Georgia Press, 1999); E. Merton Coulter, *Wormsloe: Two Centuries of a Georgia Family* (Athens: University of Georgia Press, 1955).

9. "Letter to editor," *Savannah Press*, Mar. 2, 1929 (clipping), fol. 3, box 31, GWJD; "Wormsloe Gardens," broadsheet, n.d., fol. 19, box 36, De Renne Family Papers (hereafter DFP), Hargrett.

10. "Visit! Wormsloe Gardens," pamphlet, n.d., fol. 19, box 36, DFP.

11. "Wormsloe Gardens," broadsheet, n.d., fol. 19, box 36, DFP.

12. "Garden Tours and Flower Show," pamphlet, 1935, fol. 19, box 36, DFP.

13. "Wormsloe, an Historic Plantation Dating from 1733," typescript, 1937, fol. 19, box 36, DFP.

14. "Gardens Designer Dies at 74," *Savannah Morning News* (Savannah, Ga.), Aug. 4, 1969.

15. Mr. Grossmauw to Augusta De Renne, Sept. 29, 1931, fol. "correspondence, 1931 September," box 11; David Griffiths to Augusta De Renne, Oct. 14, 1931, and O. M. Freeman to Augusta De Renne, Oct. 16, 1931, fol. "correspondence, 1931 October," box 11; B. Skeele to Augusta De Renne, Mar. 15, 1928, fol. "correspondence, 1928 March," box 9; Gardening Journal of Augusta De Renne, 1927–1929, entries throughout, fol. "Gardening material, journal," box 112; all in Wymberley Wormsloe De Renne Family Papers, Hargrett.

16. Shen Hou, *The City Natural: Garden and Forest Magazine and the Rise of American Environmentalism* (Pittsburgh: University of Pittsburgh Press, 2013), pp. 69, 73–78.

17. On this Progressive Era association and its implications, see Adam Rome, "'Political Hermaphrodites': Gender and Environmental Reform in Progressive America," *Environmental History* 11, no. 3 (July 2006): pp. 440–463.

18. Jane Brown, *Beatrix: The Gardening Life of Beatrix Jones Farrand, 1872–1959* (New York: Viking, 1995); Carolyn Merchant, "Women of the Progressive Conservation Movement: 1900–1916," *Environmental Review* 8, no. 1 (spring 1984): pp. 57–85; Joanne Seale Lawson, "Remarkable Foundations: Rose Ishbel Greely, Landscape Architect," *Washington History* 10, no. 1 (spring/summer 1998): pp. 46–69; and

Arthur Hecht, "Flowers to Gladden the City: The Takoma Horticultural Club, 1916–1971," *Records of the Columbia Historical Society*, Washington, D.C. 71/72 (1971/1972): pp. 694–696.

19. "Savannah and the Wormsloe Gardens," *Bulletin of the National Council of State Garden Clubs Incorporated* 6, no. 5 (Feb. 1936): p. 8; "Gardens Designer Dies."

20. "Wormsloe, an Historic Plantation Dating from 1733," typescript, 1937, p. 6, fol. 19, box 36, DFP.

21. Ibid., pp. 6–7.

22. "Letter to editor," *Savannah Press*, Mar. 2, 1929 (clipping), fol. 3, box 31, GWJD.

23. Visitor's Book, 1924–1938, fol. 10, box 10, DFP; "Wormsloe Gardens," *Savannah Press*, March 12, 1927 (clipping in Wormsloe Scrapbook, 1925–1929), fol. 5, box 30, GWJD.

24. "Wormsloe, Summary of Facts," memorandum, n.d., fol. 7, box 10, DFP; calculations made with worth calculator at http://www.measuringworth.com/uscompare/.

25. "Wormsloe, Summary of Facts," p. 6.

26. Swanson, *Remaking Wormsloe*, chaps. 2 and 3.

27. For what remains the best summary of these shifts, see Pete Daniel, *Breaking the Land: The Transformation of Cotton, Tobacco, and Rice Cultures since 1880* (Urbana, Ill.: University of Illinois Press, 1985).

28. Bragg, *De Renne*, pp. 254–258, 283–285; Swanson, *Remaking Wormsloe*, p. 141.

29. Bragg, *De Renne*, pp. 297–301.

30. "Influence for Good of Negro," *Savannah Morning News*, Apr. 6, 1909, p. 4, (clipping), DFP, box 36, fol. 2 (capitalization in the original).

31. Ulrich Bonnell Phillips, *American Negro Slavery: A Survey of the Supply, Employment and Control of Negro Labor as Determined by the Plantation Regime* (New York: D. Appleton, 1918); William A. Dunning, *Reconstruction, Political and Economic, 1865–1877* (New York: Harper and Brothers, 1907). On the turn of the century period of reconciliation and the memory of slavery, see David Blight, *Race and Reunion: The Civil War in American Memory* (Cambridge, Mass.: Belknap Press of Harvard University Press, 2001); Silber, *Romance of Reunion*; Gaines Foster, *Ghosts of the Confederacy: Defeat, the Lost Cause, and the Emergence of the New South* (New York: Oxford University Press, 1987), chaps. 12 and 13; Charles Reagan Wilson, *Baptized in Blood: The Religion of the Lost Cause, 1865–1920* (Athens: University of Georgia Press, 1980), chap. 8.

32. Bragg, *De Renne*, pp. 377, 387–390; Coulter, *Wormsloe*.

33. On the long history of the family's horticultural efforts, see Swanson, *Remaking Wormsloe*.

34. For the importance of "nature" in these programs, see Sarah T. Phillips, *This Land, This Nation: Conservation, Rural America, and the New Deal* (Cambridge, Mass.: Cambridge University Press, 2007); Neil Maher, *Nature's New Deal: The Civilian Conservation Corps and the Roots of the American Environmental Movement* (New York: Oxford University Press, 2008).

35. Nancy Rose, *Put to Work: The WPA and Public Employment in the Great Depression* (New York: Monthly Review, 2009); Nick Taylor, *American-Made:*

The Enduring Legacy of the WPA (New York: Bantam, 2008). The most famous of these historical endeavors was "Slave Narratives: A Folk History of Slavery in the United States," a collection of oral histories of former slaves held at the Library of Congress. The first publication of selected interviews came in the Federal Writers' Project, *These Are Our Lives* (Chapel Hill: University of North Carolina Press, 1939). Various New Deal programs created competing images of the American past. While WPA projects often conveyed nostalgia for the past, the Farm Security Administration's Historical Division, through a set of influential documentary photographs, created a compelling counter-narrative that cast many of the problems of the Great Depression as creations of poor historical practices.

36. Georgia Writer's Project, *Drums and Shadows: Survival Studies among the Georgia Coastal Negroes* (Athens: University of Georgia Press, 1940).

37. Federal Writer's Project, *Savannah* (Savannah: Review Printing Co., 1937).

38. Bragg, *De Renne*, p. 387.

39. Tammy Ingram, *Dixie Highway: Road Building and the Making of the Modern South, 1900–1930* (Chapel Hill: University of North Carolina Press, 2014).

40. Swanson, *Remaking Wormsloe*, pp. 144–146. The epitome of this elite tourism development was the Jekyll Island Club. For its history, see June Hall McCash and William Barton McCash, *The Jekyll Island Club: Southern Haven for America's Millionaires* (Athens: University of Georgia Press, 1989).

41. Elias B. Bull and Bernard Kearse, Magnolia Plantation and Gardens (Charleston County, S.C.), National Register of Historic Places nomination, South Carolina Department of Archives and History, Columbia, S.C. Available online at http://www.nationalregister.sc.gov/charleston/S10817710050/S10817710050.pdf.

42. Baird Leonard, "Mrs. Pep's Diary," *Life* 93, no. 2420 (Mar. 22, 1929): p. 12; Edward Twig, Letter to the Editor, *Forum and Century* 103, no. 3 (Mar. 1940): p. 142.

43. Swanson, *Remaking Wormsloe*, p. 137.

44. Ibid., chap. 4.

45. "Drama at Wormsloe as 'Shiners Try Steal Car," 1935 (clipping), fol. 2, box 36, DFP.

46. Swanson, *Remaking Wormsloe*, chap. 5.

47. Steven Hoelscher, "Making Place, Making Race: Performances of Whiteness in the Jim Crow South," *Annals of the Association of American Geographers* 93, no. 3 (Sept. 2003): pp. 658–659, 665–666.

48. Michele Gillespie, *Katharine and R. J. Reynolds: Partners of Fortune in the Making of the New South* (Athens: University of Georgia Press, 2012), chap. 6.

49. Tiya Miles, *The House on Diamond Hill: A Cherokee Plantation Story* (Chapel Hill: University of North Carolina Press, 2010), p. 10.

50. Anthony J. Stanonis, *Creating the Big Easy: New Orleans and the Emergence of Modern Tourism, 1918–1945* (Athens: University of Georgia Press, 2006).

51. Richard Handler and Eric Gable, *The New History in an Old Museum: Creating the Past at Colonial Williamsburg* (Durham, N.C.: Duke University Press, 1997).

52. Jennifer M. Murray, *On a Great Battlefield: The Making, Management, and Memory of Gettysburg National Military Park, 1933–2013* (Knoxville: University of Tennessee Press, 2014).

53. Mark David Spence, *Dispossessing the Wilderness: Indian Removal and the Making of the National Parks* (New York: Oxford University Press, 1999), esp. chaps. 6 and 8.

54. Katrina Powell, *"Answer at Once": Letters of Mountain Families in Shenandoah National Park, 1934–1938* (Charlottesville: University of Virginia Press, 2009); Terence Young, "False, Cheap and Degraded: When History, Economy, and Environment Collided at Cades Cove, Great Smoky Mountains National Park," *Journal of Historical Geography* 32, no. 1 (2006): pp. 169–189; Margaret Brown, *The Wild East: A Biography of the Great Smoky Mountains* (Gainesville: University Press of Florida, 2000); Justin Reich, "Re-Creating the Wilderness: Shaping Narratives and Landscapes in Shenandoah National Park," *Environmental History* 6, no. 1 (Jan. 2001): pp. 95–117; Durwood Dunn, *Cades Cove: The Life and Death of a Southern Appalachian Community, 1818–1937* (Knoxville: University of Tennessee Press, 1988).

55. On the legitimate dangers of accepting sites of historical interpretation at face value, see Seth C. Bruggeman, *Here, George Washington Was Born: Memory, Material Culture, and the Public History of a National Monument* (Athens: University of Georgia Press, 2008).

Chapter 4

"Rice Planters in their Own Right"

Northern Sportsmen and Waterfowl Management on the Santee River Plantations during the Baiting Era, 1905–1935

Matthew A. Lockhart

Historians of the U.S. Bureau of Biological Survey and its successor, the Fish and Wildlife Service, have discussed the expansion, improvement, and restoration of waterfowl habitat on national wildlife refuges in numerous books and articles.[1] This and related literature describes ecology-based waterfowl management coming of age quickly during the New Deal in an effort to rescue the continent's declining population of migratory ducks and geese, which had been decimated by decades of intensive hunting and destruction of wetlands. The number of national wildlife refuges and the total acreage in the refuge system more than doubled from 1934 to 1939. At the same time, the Biological Survey embraced the emerging science of waterfowl ecology, which stressed the role of wetland management in conservation—that is, increasing the capacity of an ecosystem to support migratory birds by artificially enhancing supplies of water, food, and cover. Expanding the refuge system and maximizing the carrying capacities of individual refuges helped to approximately quadruple waterfowl numbers between 1935 and 1944. This dramatic recovery of waterfowl populations is one of the greatest success stories in the history of American wildlife conservation.[2]

Many may find it surprising to learn that another pillar of waterfowl conservation in the first half of the twentieth century were duck-hunting clubs, which acquired vulnerable wetlands, protected them from drainage, and in the process, added critical links to the chain of habitat provided by national wildlife refuges. An estimated 6,000 ducking clubs in forty-eight states controlled roughly 3 million acres in 1963. By comparison, in the same year, the 220 national wildlife refuges managed primarily for waterfowl comprised slightly less than 2.6 million acres. Louisiana, the heart of the Mississippi

Flyway wintering grounds, and California, winter home to the majority of waterfowl in the Pacific Flyway, had the most clubs of any of the states with around 1,000 each. Louisiana had the most privately managed acres, 1.5 million. Clubs managed the most acres per capita in South Carolina, where fourteen had possession of a total of 70,000 acres.[3]

While the rise of waterfowl management on public wetlands has received extensive study, historians have paid little attention to the private wetlands held by hunters. The history of private citizens manipulating large areas of wetlands out of concern for waterfowl shooting and conservation is actually much older than the federal refuge system. Duck shooting for pleasure became popular during the nineteenth century, largely because of the growth of the urban middle and upper classes. From an early date, duck hunters organized clubs, pooled their financial resources, bought and leased choice wetland tracts as shooting preserves, and incorporated habitat management in their hunting strategy. Knowledge of their preferred quarry, especially its mobility and habit of congregating in small areas, inspired their efforts, which predated similar measures taken by sportsmen who pursued upland game.[4] Duck hunters' activities raise a powerful question: what did waterfowl management mean before the age of ecology?

Many of the oldest ducking clubs in the United States lay along the Chesapeake Bay, which attracted large numbers of wintering waterfowl and hordes of sport and market hunters. By the 1850s, some sportsmen sought out less-crowded locations such as the Currituck Sound in North Carolina and Ohio's Lake Erie marshes. After the Civil War, with improved shotguns adding to the popularity of wing shooting, upstart ducking clubs gained control of almost all of the wetlands convenient to major cities in the Northeast, the Midwest, and northern California. As the national transportation network rapidly expanded, gentleman duck hunters finally reached the southernmost wintering grounds in the continental United States—including the South Carolina lowcountry, the Gulf coast of Louisiana and Texas, and central and southern California—at the end of the century. By this time, overhunting and the inroads of agriculture on the northern prairie breeding grounds had sent waterfowl populations plummeting. In a new era of more hunters and fewer ducks, sportsmen had no choice but to try to lure birds within range of their guns (and out of range of someone else's). They did so by impounding wetlands, manipulating water levels, and propagating native marsh plants like northern wild rice (*Zizania palustris*), wild celery (*Vallisneria americana*), and sago pondweed (*Stuckenia pectinata*) as duck food. The San Joaquin Shooting Club on Newport Bay in Orange County, California, became one of the first to erect impoundments. "All this work improves the preserves by enlarging the water surface and opening new feeding grounds for the ducks," the *Los Angeles Times* observed in 1896.[5] Without understanding it as such, sportsmen began increasing the carrying capacities of their properties. Habitat management on private ducking preserves preceded similar work on national wildlife refuges by several decades.[6]

Regulating water levels played a crucial role in creating artificial duck marshes. Sportsmen generally built dikes, dug ditches, and installed floodgates for this purpose. The situation differed in the lowcountry region of South Carolina, however, where established patterns of wetlands use presented unique management challenges. In other parts of the country, sportsmen developed duck marshes in estuaries and inlets that were, for all practical purposes, ecologically pristine. In South Carolina, they encountered wetland environments that had supported intensive agriculture for centuries. Whereas other duck hunters constructed impoundments on their preserves one by one from the ground up, those who moved into South Carolina inherited a coastline already checkered with thousands of them.

Throughout the eighteenth century and for the better part of the nineteenth, South Carolina was a global leader in commercial rice production. Rice plantations dominated the coastal landscape. From the mid-eighteenth century onward, South Carolina planters grew rice in fields along tidal rivers. Mile after mile of intersecting dikes, ditches, and canals punctuated by culverts and gates allowed for precise control of water levels. By the 1850s, when development of the state's rice infrastructure reached its zenith, clusters of plantations formed immense impoundment grids at intervals from the Waccamaw River north of Georgetown to the Savannah River on the South Carolina-Georgia border, a distance of approximately 160 linear miles.

During the last decade of the nineteenth century and first quarter of the twentieth, scores of South Carolina rice plantations, once some of the most valuable agricultural land in the world, fell idle. The decline of commercial rice production and planters' failure to find alternative uses for their lands dealt a crushing blow to the agricultural system that had historically sustained the coastal economy. Landowners eagerly sold and leased large tracts to wealthy sportsmen from the North, who used the century-old rice fields as duck-shooting preserves. More differences than similarities existed between prototypically modern ducking preserves, which concentrated on growing wild marsh plants in artificial impoundments, and those established on lowcountry rice plantations. From the 1890s until the eve of World War II, managing wetlands and waterfowl in the lowcountry chiefly meant sowing, tending, and reaping rice in a manner no different than Carolinians had practiced for generations, save for the milling and marketing of the grain. Ducks consumed most of the rice planted by sportsmen in rough form where it had been strategically scattered as bait. Historians who have studied the end of rice planting in South Carolina approach the topic strictly from the standpoint of commercial production.[7] In fact, a number of old plantations continued to produce rice for years, sometimes decades, after the last crop was sold. The Santee Club operated the largest and longest-running rice-planting operation on a ducking preserve in South Carolina. Examining the club's activities demands that sportsmen be seen as ecological actors in the pre-ecology era and illuminates a chapter of the history of lowcountry rice culture that historians have virtually ignored.

The Santee Club was chartered in 1898 with eleven members, but for most of its existence—from 1900 until 1974, when the club transferred the title for its vast hunting preserve to the Nature Conservancy and effectively disbanded—membership shares were capped at forty. Over the years, the nearly two hundred individuals who belonged to the Santee Club were among the wealthiest and most privileged sportsmen in America. Their ranks were filled with bankers, capitalists, corporate attorneys, stockbrokers, industrialists, entrepreneurs, and heirs to Gilded Age fortunes—the cream of the eastern elites from Massachusetts, Connecticut, New York, New Jersey, Pennsylvania, Delaware, and Maryland.[8] As early as June 3, 1899, the day after millionaire drug manufacturer Isaac E. Emerson received his certificate of membership, the *Baltimore American* opined that the Santee Club was "the most influential gunning club in the United States."[9] Clarence H. Mackay, head of his father's international cable-telegraph empire, became a member of the club the next year, as did second-generation department-store magnate Eben D. Jordan, Jr.[10] Jordan's son Robert and the Hoyt brothers—Edward, Theodore, Walter, and George, heirs to their father's interest in the United States Leather Company—became members by 1906.[11] Steel tycoon Andrew Carnegie's nephew Frank was elected to membership five years later; William L. McLean, publisher of the Philadelphia *Evening Bulletin*, followed in 1913; and the son of famous financier Jay Cooke joined in 1925.[12]

Before the middle of the next decade, the presidents or board chairmen of the Standard Oil Company, the Penn Mutual Life Insurance Company, the Corn Exchange National Bank and Trust Company, the Pennsylvania Railroad, the Fidelity-Philadelphia Trust Company, the Chemical Bank and Trust Company, and the Lehigh Coal and Navigation Company had joined the Santee Club. So too did Edward Hoyt's son Oliver, William McLean's sons Robert and William, Jr., and Jay Cooke III.[13] "There probably is no similar organization in the United States that has such a group of men high in the world of large interests," wrote journalist John Vavasour Noel after visiting the club in 1932.[14] Grover Cleveland, the twenty-second and twenty-fourth president of the United States, held an honorary membership in the club up until his death in 1908.[15]

The Santee Club distinguished itself not only by the influence and affluence of its members but also by the extent of its land. During the club's heyday, it controlled twelve former rice plantations at the mouth of the Santee

River in Charleston and Georgetown counties. Together, these lands totaled approximately 25,000 acres, with all but about a quarter in wetlands. Spanning plantation fields, miles of tidal marshes, and several barrier islands, its duck-shooting preserve dwarfed most others in South Carolina, which rarely comprised even half as many plantations, and ranked among the largest in the country, equaled in area only by a few of the San Joaquin River preserves in California's vast Central Valley.[16]

By the time the Santee Club came on the scene, South Carolinians had been impounding tidal wetlands to grow rice on large plantations for over 150 years. Tidal rice cultivation relied on the natural rise and fall of the ocean tides to irrigate and drain fields in river floodplains. This technique could only be applied to the zone of tidal influence above the upstream limit of saltwater incursion, which varied from within a few miles of the open ocean to as many as forty miles inland. The suitability of a site for tidal culture depended on where a river originated (in the piedmont or on the coastal plain), its volume of freshwater, the size and shape of its estuary, and additional hydrological factors.[17]

Twice a day as the tide rose, a wedge of dense saltwater entered the mouths of the rivers, displacing freshwater and forcing it back upstream. At high tide, this action raised the water level in the lower reaches of the rivers and their tributaries anywhere from a foot or so at some locations up to ten feet at others during spring tides. On the ebb tide, the level of the water dropped and the normal flow of the rivers resumed.[18]

During the second half of the eighteenth century, planters tasked slaves with reclaiming thousands of acres of tidal wetlands and developing an intricate system of dikes, canals, ditches, and ingenious tide-operated, gravity-flow water-control structures called "trunks"—rectangular wooden culverts constructed of heavy cypress lumber and fitted with a hanging gate at each end. Such infrastructure allowed planters to use predictable fluctuations in water levels for flooding and draining rice fields. Tidal rice production generated enormous wealth for the plantation owners. It also made extensive alterations to the wetlandscape. The Herculean task of creating and maintaining tidal rice plantations by hand led planters to invest considerable capital in armies of enslaved laborers. Planters preferred slaves from rice-growing regions of West Africa that used tidal culture. Between 1750 and 1770, South Carolina's slave population more than doubled, and in parts of the rice belt, blacks came to outnumber whites as much as nine to one.[19]

Wresting rice fields from tidal marshes and swamps required back-breaking labor. Slaves carried out their work entirely by hand, using axes, spades, and hoes as tools. The first step was to put up a large earthen dike, usually referred to in South Carolina as a "bank," along the river's low-water line that was substantial enough not to be overflowed by the spring tides. Banks typically stood about five feet high and measured three feet across at the top and twelve to fifteen feet at the base. Sloping sides reduced erosion. Some banks followed the contours of the river for miles, enclosing hundreds of acres. Next, a ditch was excavated between the bank and the field, and a trunk running from the ditch to the river was buried in the bank. The field was then cleared, leveled, and divided by "cross" banks into sections of ten to forty acres, which were further subdivided into rows of planting beds. Each section, or "square," was a separately functioning impoundment with shallow internal ditches known as "quarter drains" and a trunk for precise regulation of the water. All were connected back to the main ditch by a network of "face" ditches running between the squares. Following completion of the field fronting the river, others could be built behind it and supplied with water directly from the river by means of a canal skirting the original embankment. Reclaiming a single rice field required slaves to move tons upon tons of earth.[20]

With this elaborate irrigation system in place, flooding a rice crop was simply a matter of raising the gate on the end of the main trunk closest to the river when the tide was higher than field level. River water poured through the trunk, pushing open the interior gate and flowing into the field. Pressure exerted on the inside gate when the tide ebbed forced it to shut, preventing the water in the field from flowing back through the trunk and into the river. After a succession of high tides, the entire field was inundated with freshwater. Reversing the process at successive low tides left the field dry.[21]

Planters employed tidal technology on rivers all along the Carolina coast, including the Cooper, the Ashley, the Edisto, the Ashepoo, the Combahee, and the Savannah. Plantations on the Waccamaw, the Great Pee Dee, the Little Pee Dee, the Black, the Sampit, and the Santee rivers in the Georgetown District ultimately became the most productive in the state. At their peak around 1850, South Carolina plantations produced three-quarters of the rice grown in the United States, upward of 150 million pounds per year. The Georgetown District produced a little less than half of the state's total. During the peak of commercial rice production, rice impoundments on lowcountry plantations encompassed approximately 150,000 acres, or nearly 30 percent of all tidal wetlands in the state of South Carolina.[22]

After the Civil War, South Carolina's tidal plantations entered a period of protracted decline. The rice industry survived the physical damage of the war, concomitant disruptions to landholding and capital, and the transition to free labor, but planting never again approached its antebellum scale. Rice production plummeted in the late 1860s—from 119,100,528 pounds in 1859 to 32,304,825 in 1869—and then slowly recovered in fits and starts over the next three decades. Yet the state's postbellum rice crop only surpassed 50 million pounds once, in 1879, and came close again in 1899. Planters grew rice on 78,388 acres in 1880, about half as many as before the war.

In the 1880s, South Carolina lost its position as the leading rice producer in the nation to Louisiana, where the industry gained a competitive advantage through corporate financing, mechanization, and a new style of prairie farming. Increasing competition from producers in Louisiana, Texas, Arkansas, and Asia and a series of hurricanes between 1893 and 1911 forced many South Carolina rice planters out of business. The end of commercial rice production opened the door for sportsmen from the North to lease and purchase devalued rice plantations as duck-shooting preserves.[23]

Rice plantations possessed extraordinary complexity. A journalist who visited South Carolina during Reconstruction famously referred to a working rice plantation as "a huge hydraulic machine, maintained by constant warring against the rivers," and one prominent planter attested that "the whole apparatus of levels, floodgates, canals, banks, and ditches is of the most extensive kind, requiring skill and unity of purpose to keep in order."[24] In addition to their environmental intricacies, rice plantations embodied complex patterns of historical development and fraught social relations. By the late nineteenth century, many plantations had been abandoned for years, which resulted in deterioration of the expensive impoundment infrastructure. Thus, the most desirable plantations for duck shooting were those still "under bank"—in working order—and, therefore, populated by communities of former slaves and their descendants whose ties to the land went back multiple generations.

Each year from late October through March, millions of migrating waterfowl flocked to rice impoundments, which planters customarily flooded after harvest and left littered with waste grain. The combination of abundant standing water, ample food, and a mild climate led author Archibald Rutledge, who grew up on a Santee River rice plantation, to call coastal South Carolina "a regular Riviera for wintering wild fowl."[25] For all intents and purposes, well-maintained plantations provided turnkey ducking preserves. The challenge for the new owners lay in keeping them that way as the transition to recreational use occurred. "Soon," wrote journalist Chalmers S. Murray, who frequently reported on what he privately called "the second invasion" for the Charleston *News and Courier* in the early 1930s, "the new owners became rice planters in their own right."[26]

Although the Santee Club's 1898 constitution allowed for "rais[ing] such plantation, farm and garden products upon the real estate owned by the Club, as the Club may desire," members did not start out to become rice planters.[27] A little over a year after the constitution was adopted, Charleston attorney Theodore D. Jervey, Jr., who had brokered the deal for one of the club's keystone tracts on the South Santee River, remarked, "In purchasing Blake's [Plantation] to a great extent, they got a white elephant; for they have no desire to cultivate the land and that is where its value lies."[28] In 1850 thirty-nine Santee River plantations had grown rice on 16,660 acres, representing practically the entire delta from its head, where the North Santee and South Santee diverged about fifteen miles inland, to the marshes of South Island, Cedar Island, and Murphy's Island, where the rivers emptied into the Atlantic Ocean.[29] A half-century later, rice fields still covered extensive portions of the delta. "In 1900 and for many years after, there were very large rice plantations on both Santee rivers," recalled Santee Club member Henry H. Carter, a civil engineer from Boston who owned his own contracting company.[30] Speaking specifically of the South Santee, which formed the spine of the club's holdings, Carter stated: "Beginning at about our Club House [at Blake's Plantation, approximately five miles west of the islands] on the south side of the river, the rice fields extended up river [*sic*] for miles. Messrs. Doar, Lucas, Lowndes, Rutledge, Seabrooke, Graham and many others raised rice. On the north side of the river Mr. Beckman and others raised rice at Blackwood, Fanny Meade, Tranquility and many other plantations."[31]

Santee Club members shot ducks mainly over shallow brackish ponds sheltered in natural depressions of the salt marshes adjacent to Cedar Island, Murphy's Island, and Little Murphy's Island. In the early years, the sport depended entirely on the proximity of the rice plantations. Each day, ducks wintering on the Santee Delta, mostly mallards (*Anas platyrhynchos*), migrated en masse from nighttime feeding grounds on delta plantations to daytime resting grounds on the coast and back again. "Thousands and thousands of ducks passed down the river at daybreak and returned up river at night," related Carter. Gunners got to their stands in the marsh before dawn, and, according to Carter, "It was seldom one saw any ducks on entering a pond."[32] Once the morning "flight" commenced, pairs and small flocks of ducks descended on the ponds, one after another, for several hours. Shooting at Murphy's Island, Carter brought down 152 ducks on the morning of November 21, 1903. A notation in the Santee Club's shooting log observed, "There was a splendid flight at Point Stand [Pond] and at 10:30, when Carter's shells gave out, the mallards were still 'pouring in.'"[33] After the morning flight ceased, Carter noted that one could expect "Ocean Pond, Black Point, Beach Pond, Wood's Pond, Graveyard, Coy, Peter and many others" to be "crowded with ducks during the day."[34] On November 25, 1904, the steamboat carrying Carter and four other Santee Club members and their guides to Murphy's Island broke down two miles from the wharf, forcing the party to take the rowboat the rest of the way. "This made the crowd late," one of them wrote in the log, "and the ducks were all in Black Point to the extent of 100,000 or more."[35]

Until the mid-1880s, most of the Santee Delta remained planted in rice. The largest crop since the war lay ripening in the fields when two devastating hurricanes struck the region within seven weeks of each other in 1893. By 1907 the total acreage on the North and South Santees devoted to rice had dwindled to 1,400. The last substantial crop was planted the next year, and a September freshet destroyed much of it only a few weeks before harvest time.[36] As planters abandoned plantation after plantation on the delta over the next few years, the Santee Club faced the prospect of having to manage waterfowl habitat itself.

The Santee Club took an initial step toward planting its own rice by hiring Ludwig A. Beckman as its first full-time manager on April 15, 1905. Beckman took the job after selling his plantation in the delta, Blackwood, to club member Eben Jordan of Boston, who used it as a personal shooting ground during his visits to the club.[37] Although it cannot be corroborated, Beckman may have overseen planting of the club's first rice crop within a couple of months of being hired. A small collection of Beckman's personal and business papers suggests that the club planted a crop in 1906. An expense sheet from June of that year shows the club purchased twenty-five bushels of seed rice from Edward Porter Alexander, a former Confederate general and railroad executive from Georgia who had taken up rice planting at Ford's Point Plantation on South Island in retirement.[38] Twenty-five bushels would have been enough seed to plant about ten acres of rice.[39] Beckman also left behind a number of journals that are held in a private collection, but the oldest surviving volume goes back no further than 1919. Aside from these sources, the only detailed evidence of rice cultivation at the club comes from contemporaneous newspapers.

In a letter to the editor of the *News and Courier* published in January 1908, Beckman stated that the club's "old rice lands are being reclaimed and planted" by "hundreds of day laborers."[40] A Charleston *Evening Post* article from March 1911 indicated that the club's early reclamation and planting efforts had met with success and were ongoing: "During the summer months, thousands of dollars are spent on improving their property for the next shooting season. The club will plant a large area in rice this year in order to attract the ducks next winter."[41] As nearly as can be determined, only one hundred acres of rice were planted on the South Santee in 1916, and the club was responsible for seventy-five of them.[42] An article in the *News and Courier* the following year reported that "the club plants seventy acres of rice, solely for duck feed."[43] The club increased its acreage in rice to one hundred in 1918, and after the 1919 growing season, during which a large part of the crop was lost to a mid-summer freshet, it invited proposals from contractors for a major expansion of its planting operations. An advertisement published in the *Evening Post* explained, "The Santee Club is planning to reclaim 500 acres of old rice fields on the Santee River to be planted in rice and would like to get bids from dredging concerns on opening up canals, building dykes, etc." Judging from Beckman's journals, the club did not follow through on this plan and probably never planted much more than one hundred acres of rice in a season.[44]

As member B. Brannan Reath II of Easton, Maryland, put it, the Santee Club grew rice "in the old-fashioned way" throughout the 1920s and 1930s.[45] Until 1939, when workers first used a tractor instead of oxen to harrow one of the fields, all of the work was accomplished by hand, with most of the laborers coming from the nearby African American community of Collins Creek.[46] There were two windows for planting rice in South Carolina: one in April and a second in June. Beckman's journals reveal that the Santee Club invariably planted during late May or June. The club's top field hands also worked as guides and watchmen during the winter, so they got a late start in preparing the rice fields for the growing season. Cleaning ditches took place mostly in March and April, followed by plowing in May and disking in early June. Women joined the club's summer work force. They "clayed" the seeds with marsh mud to prevent them from floating when water flooded the fields and then assisted with the planting, especially in low places that stayed too soggy for sowing with a seed drill pulled by an ox or a mule. Women carried out much of the hoeing in July and August while men built and buried new trunks and repaired and strengthened the banks of fields left fallow for maintenance. Upkeep of the banks and trunks involved year-round work, except during the ducking season. In late summer, Beckman deployed men and boys as "bird minders" to protect the ripening grain from bobolinks (*Dolichonyx oryzivorus*) and red-winged black birds (*Agelaius phoeniceus*). All hands, men and women, took to the fields in October or early November to cut, stack, and tie the rice. Women threshed and winnowed it overwinter in the club's "rice yard." Beckman often did not supervise the plantation work directly, but instead used the traditional lowcountry "task system," whereby hands performed specified assignments per day, at their own pace, and earned a day's pay upon completion of the "task." For example, cleaning a one-acre length of "big ditches" represented one task in April 1924, as did cleaning a two-acre length of "small ditches." Each of these tasks paid one dollar.[47] Beckman usually handled the delicate business of irrigation and drainage himself, "flowing" and "running off" the fields at the appropriate intervals during the cultivation cycle.

Along with paying weekly wages, the Santee Club offered field hands a share of the rice crop. From planting to hoeing to cutting to winnowing, the "club rice" always received first consideration. Once it had been tended, workers turned their attention to the "share rice." Beckman wrote in 1919 that this arrangement "enabl[ed] people to produce rice for food who otherwise could not without the club's assistance."[48] Everyone connected with the Santee Club ate rice from its fields. The hundreds of African Americans employed by the club subsisted on their share of the harvest, while the kitchen staff at the clubhouse daily served rice from the club's share to visiting parties of hunters. The rest of the club's portion went for next year's seed and baiting the marsh ponds.

For recreational hunters and professional hunters alike, shooting ducks over bait was a tried-and-true method for success. "The principal involved in baiting is relatively simple," writes modern-day Maryland waterfowler Harry M. Walsh. As he explains, "Bait is placed in a convenient spot until it has been discovered by the waterfowl. Their numbers then become a simple ratio to the amount of bait. Once the flight and feeding pattern has been established, good hunting is assured. The ducks can then be conditioned to feed when and where hunters desire."[49] Ducks responded to a variety of cereal grains, so hunters enticed them with whatever kind they could obtain most economically. The vast majority of hunters used corn.

Duck baiting originated on the Chesapeake Bay. Some of the oldest documentation of baiting on the Chesapeake dates to 1892, though it doubtless started earlier. In March of that year, twelve months from being inaugurated as president of the United States for the second time, Grover Cleveland shot ducks as a guest of the Spesutia Island Rod and Gun Club, located on the headwaters of the bay in Harford County, Maryland, near Havre de Grace. The ex-president was an avid outdoorsman, and his "luck" on fishing and hunting trips was often the subject of national news. A correspondent of *Forest and Stream*, the country's foremost sporting magazine, disclosed that Cleveland's luck at Spesutia Island had a lot to do with baiting. "It is the club's practice to bait their blinds, putting out twenty bushels of corn at a time," explained the *Forest and Stream* article, which the *New York Times* reprinted two days later.[50] A Baltimore *Sun* article on the Spesutia Island Rod and Gun Club from December 1894 stated that "every season the club puts out hundreds of bushels of corn at different places about their points and marshes."[51]

Before the decade was out, duck-hunting clubs on the Chesapeake Bay used baiting extensively. "Much competition occurs among the proprietors of the shooting shores," the *Sun* declared. "During recent years baiting has become necessary to hold the stock of ducks at the ponds," noted the newspaper, "and many hundred bushels of corn will have been consumed before the shooting begins." This led the *Sun* to conclude that "the baiting system" was "the most expensive factor of modern ducking on these marshes."[52]

Another example of baiting in the 1890s comes from South Carolina and also involves Grover Cleveland. Railroader-turned-planter Edward Porter Alexander baited ponds on South Island with rice from his plantation as early as January 1893.[53] An accommodating host whose connections and hospitality made him the most influential gentleman waterfowler in the state during the nineteenth century, Alexander welcomed a parade of dignitaries to hunt ducks on his property. Over four and half years after Alexander initially extended the invitation, Cleveland finally made his first trip to South Island. Cleveland liked the shooting and Alexander's company so much that he came

back eleven times in the next thirteen years. Little more than acquaintances
in the beginning, the two men became close friends.[54] Journalist James Henry
Rice, Jr., who also shot ducks at South Island as Alexander's guest, claimed
that his host "has two ponds which he keeps for the exclusive use of his
friend, Mr. Cleveland, and which ponds are baited daily for months, or as
often as it is necessary."[55] If this was true, then Alexander never admitted it,
not even in his own journal.

Cleveland shot at the Santee Club on a number of occasions while staying
with Alexander—enough to be named an honorary member of the club in
1900—but did so just once, in March 1907, after the club hired Beckman.[56]
The club's acreage in rice was small then, and that late in the season, all of
the bait grown in 1906 may have been gone. Whatever the reason, there is no
record of a pond ever having been baited for Cleveland at Santee.

In later years, each guide at the Santee Club took a bushel of rice in his boat
to the blind in the morning. When the club member or guest finished shooting, the guide scattered the rice on the surface of the water near the blind.
In the afternoon, a special crew would make the rounds to the blinds that
had not been occupied that day and bait them. "This sometimes meant that
bait would be put out in forty areas," Reath observed.[57] Beckman might put
out even more while making inspections of the property, such as on January
20, 1928, when he baited the old fields at his former plantation, Blackwood
(the Santee Club had bought it from the Jordan estate three years earlier).[58]
Considering that a bushel of rough rice weighed forty pounds or more and
allowing for "rest days," when no shooting took place, the club put out as
much as fifty tons of rice over the course of a November-to-March season.[59]
This matched the larger Chesapeake Bay clubs, which used between forty
and 100 tons of corn every year.[60] In addition to baiting with rice versus corn,
another major distinction between the Santee Club and most of those on the
Chesapeake was that the latter bought grain, while the former grew it.

By heavily baiting its ponds with rice, the Santee Club gradually altered
the flight and feeding patterns of the ducks on the Santee Delta. Over time,
ducks started spending day and night in the Santee Club marshes. "Ducks still
feed up river but to a less extent," Carter remarked.[61] As a result of operating its own rice plantation and baiting aggressively, the club actually bagged
record numbers of ducks in the 1920s, well after the demise of commercial
rice planting on the Santee River. In 1904–1905, the last season before Beckman began as manager, the club took a total of 3,613 ducks.[62] Even though
the loss of breeding habitat and overhunting caused fewer ducks to return to
the lowcountry with each passing season, the club's total for 1922–1923 was
6,388.[63] The 77 percent increase in the number of ducks killed proved all the
more impressive because it followed passage of the Migratory Bird Treaty
Act in 1918, which empowered the federal government to regulate waterfowl

shooting for the first time by shortening the open season and reducing the daily bag limit.

By 1931–1932, with the open season on ducks cut to thirty days and the daily bag limit lowered to fifteen, the Santee Club did more feeding than baiting. What once had been a means of ensuring the biggest bags possible was now primarily seen as a management and conservation tool. Beckman started scattering rice for the ducks as soon as they arrived in the fall and persisted, to a greater or lesser degree, until they left in the spring. In late February 1932, more than two months after the season closed—during which club members killed 1,831 ducks—a team consisting of staff from the Charleston Museum and Cape Romain Migratory Bird Refuge conducted a duck census at the club.[64] "To this small group," wrote museum ornithologist Alexander Sprunt, Jr., "the day was a revelation!" After Sprunt and his colleagues surveyed roughly one-third of the wetlands on the property, "the result was that, at the most conservative estimate, it was decided that sixty-nine thousand ducks had been observed!" A fraction of the club's marshes and rice fields holding this many ducks "seems rather remarkable in this day and time when the numbers at large seem to have decreased so much," Sprunt confessed, adding that "the Santee Gun club [*sic*] is run along lines approaching perfection."[65]

In 1935 new federal regulations enacted under the authority of the Migratory Bird Treaty Act outlawed the practice of shooting over bait.[66] The new rule proved so consequential for the Santee Club that Beckman actually attempted to contact Jay N. "Ding" Darling, chief of the U.S. Bureau of Biological Survey, a few weeks before the opening of the 1935–1936 season seeking clarification of the law. "I have been studying the situation," Beckman wrote, "and I am afraid that I could not feed the ducks on the Club property during the shooting season, and not have some of the ducks moving over some of our ponds where there is shooting, while they are going to and from their feed." He asked if it would be within the law for the club to stop feeding the ducks one day prior to the opening of the season on November 20, or even four days prior on the 16th, then resume the day after the season ended on December 19. Beckman closed, "I understand some are doing this, but I will not, until I am sure it is permissible."[67] The bureau's response, which came from Stanley P. Young, chief of the Division of Game Management, put an end to the baiting era at the Santee Club: "The regulation, as you doubtless realize, forbids the shooting of migratory waterfowl attracted to the hunter with or by aid of feed. Now, at what time the feed is put out is immaterial if there is a direct connection between the feed and the ducks that are shot."[68]

Beckman's letter alludes to others on the Santee River feeding ducks. The Santee Club may have left behind the most thorough documentation of its activities related to plantation-based waterfowl management, but it did not act alone. Scattered sources give a sense of the scope of the rice planting and baiting carried out by numerous clubs and individual preserve owners between the World Wars. Next in order after the Santee Club was the Kinloch Gun Club. Founded in 1912 by a group of sportsmen from Wilmington, Delaware, with ties to the DuPont chemical company, Kinloch occupied 8,000 acres encompassing nine plantations on the North Santee.[69] Limited documentary and photographic evidence and articles written in the 1930s by Charleston *News and Courier* reporters Chalmers Murray and John M. Lofton, Jr., indicate that the Kinloch Gun Club planted rice from 1913 to at least 1938.[70] Like the Santee Club, Kinloch employed retired white rice planters to manage its preserve, and scores of African American laborers tended the rice crops. In the 1910s, its rice acreage equaled, if not surpassed, that of the Santee Club, and Kinloch even milled and sold some of its excess grain on the open market, making it one of the last commercial growers on the delta.[71] Moreover, records show that the Santee Club and the Kinloch Gun Club sold seed rice to neighboring sportsmen. Examples include Idaho rancher and mining engineer Wayne Darlington, who produced rice crops at Annandale Plantation in 1918 and 1919 with Kinloch seed; New Yorker E. Gerry Chadwick, a real-estate executive and member of the Santee Club, who used fifty bushels of Santee seed rice to plant the Wedge, his private plantation, in 1933; and former cement-company president William N. Beach of New York, who likewise obtained fifty bushels of seed rice from Santee in 1938 and again in 1939 for his plantation, Rice Hope.[72] Besides the Kinloch Gun Club, Murray wrote articles on Annandale, the Wedge, Rice Hope, and several other Santee River plantations in 1931.[73] Reflecting eighteen years later on the people and places he had covered in the stories, Murray asserted that "almost all of the millionaire sportsmen followed the same pattern": in addition to turning upland fields into hunting grounds and restoring the old mansions, they "grew rice for ducks."[74] Waldo L. McAtee, acting chief of the Biological Survey, confirmed the extent of rice growing and its importance to waterfowl shooting and management on the Santee Delta in 1931, four years before the advent of the federal baiting ban. "I have been all over the property of the Santee Club and we have available reports of a special investigator who covered most of the club properties in the region," wrote McAtee, a specialist on the feeding habits of migratory waterfowl. He concluded that "essentially the ducking properties of the lower Santee region are kept going by baiting."[75]

    As the Rice Hope Plantation example illustrates, rice culture continued at the Santee Club, the Kinloch Gun Club, and other places even after the baiting ban took effect. The most detailed timeline of the closing days of rice planting on a Santee River ducking preserve comes from the Santee Club, where it endured until the early 1940s, nearly a decade and a half after the last commercial rice grower in South Carolina, Theodore Ravenel of Laurel Spring Plantation, gathered his final crop on the Combahee River southwest of Charleston.[76] Historian James H. Tuten contends that Ravenel and others of his generation persevered in the industry despite environmental and economic difficulties because planting rice in the lowcountry "involved culturally defined self-identity as much as the desire to make money."[77] After growing rice for thirty years, the Santee Club may have persisted after the prohibition on baiting partly out of a similar sense of self-identity. Some rice was put out for the ducks after the shooting seasons closed, but most of the grain grown at the Santee Club in the late 1930s was either eaten by members and employees or saved for seed. Because of outmigration and better job opportunities elsewhere, field hands had become scarce. Increasing labor costs caused the club to scale back its planting activities. It grew its last rice crop in a single, thirty-acre field in 1941.[78]

In April 1942, the South Carolina Public Service Authority completed construction of a dam across the Santee River at Wilson's Landing, part of the New Deal-funded Santee-Cooper Project. Designed to produce hydroelectric power and develop inland navigation, the project diverted 90 percent of the Santee's freshwater, the fourth largest average flow by volume of any river on the Atlantic coast of the United States, into Charleston Harbor via the Cooper River.[79] The Santee Club, the owners of other ducking properties on the lower Santee, the U.S. Bureau of Biological Survey, and a host of environmental organizations had opposed the Wilson's Landing Dam, but to no avail. With its primary source of vital freshwater all but eliminated, Beckman jotted in his journal that the club "cannot plant rice anymore."[80]

Back during the heady days of the 1920s, the members of the Santee Club had recognized that planting rice would only become more expensive in the future. In 1921, on the heels of considering reclaiming hundreds of additional acres of old rice fields, the club brought in the first of several outside consultants—ranging from a North Carolina duck-hunting guide to aquatic nurserymen from Wisconsin to government scientists—to instruct Beckman on identifying and propagating native perennial marsh plants such as widgeon grass (*Ruppia maritima*) and nut grass (*Cyperus esculentus*) that could be easily transplanted and supply the club with a permanent source of duck food in the ponds at less expense than baiting with rice.[81] By the time of the baiting ban, the club's program in propagating native duck foods was well established, and its distinctive style of managing its preserve for waterfowl, a product of the agriculturally depressed lowcountry of the early twentieth century, had fallen into step with the national mainstream. As monoculture gave way to biologically diverse duck marshes in the years leading up to the ban on baiting, the ecology of the club's preserve grew increasingly similar to the modern prototype.

In his writings, Archibald Rutledge often depicted the Santee Delta during his lifetime as a "wilderness." In *Home by the River* from 1941, he described "the lonely delta of the Santee, formerly one of the greatest rice-growing areas of North America, but now returned to a green wilderness as primeval as it must have been in the days of the Indians."[82] Rutledge evoked the image of a reedy wasteland—vast, unbroken, and "as primeval as it must have been in the days of the Indians"—to lend his stories and poems an element of romance and mystery, but he took a measure of creative license in the picture he painted. Thousands of acres of delta rice fields had been abandoned, and countless miles of banks had degraded since the Civil War. But even as Rutledge wrote *Home by the River*, rice grew here and there along the North and South Santees in well-ordered impoundments that had been maintained over the course of many years, including more than a few at Blake's Plantation on the Santee Club grounds—lands that his father had once planted.[83] "I have, as a cherished recollection [from childhood], the vision of a glorious field of a thousand acres of rice, level and golden, stretching between the two broad rivers toward the sea," Rutledge wrote.[84] Several hundred acres of rice dotting the delta wetlandscape in the 1930s paled in comparison to what Rutledge had witnessed as a boy, but it is nonetheless significant, especially considering the late date and who did the planting.

Northern duck hunters extended the life of working rice plantations on the Santee River by some two to three decades, repurposing age-old wetland-use customs and investing large sums of capital to prop up a dying culture for a little longer so that they might enjoy their favorite sport to the fullest. Accounts in contemporary newspapers and the records of the Santee Club, the Kinloch Gun Club, and the Biological Survey indicate that sportsmen on other lowcountry rivers did the same. In 1906 Santee Club member Isaac Emerson began acquiring a string of plantations on the Waccamaw River for his personal use that eventually encompassed almost 10,000 acres.[85] When he arrived from Baltimore for his first hunt of the 1909–1910 season, a *News and Courier* article announced that "Mr. Emerson has had his extensive rice lands flooded for the duck shooting and baited it with uncut rice. . . . It is said that even the blinds are formed of sheaves of rice."[86] Author Elizabeth Allston Pringle took note after World War I that Emerson continued to have "some fields planted in rice every year, simply for the ducks."[87] Emerson's neighbor on the Waccamaw, New York financier Bernard M. Baruch, purchased nine contiguous plantations totaling nearly 12,000 acres between 1905 and 1907. Baruch bought 115 bushels of seed rice from the Kinloch Gun Club in 1919. When he acquired the first six of his plantations from the Donaldson family in 1905, they contained 225 acres of rice fields in a "high state of cultivation."[88] Fourteen years later, it seems that Baruch had plans to plant no less than fifty acres or so. A tantalizing clue about the extent of rice planting across the lowcountry region in the mid-1930s comes from Neal Hotchkiss, a federal wildlife biologist who surveyed waterfowl conditions in South Carolina in

December 1934, in the final year before baiting became illegal. Hotchkiss found that "the amount of baiting appeared to be as great as a year ago" not only on the Santee River but also on the Cooper and the Combahee, and "it may have been heavier in proportion to the numbers of ducks present."[89] Three years after Hotchkiss made his report, the owner of Mansfield Plantation on the Black River, a Philadelphia stockbroker, ordered forty-four bushels of seed rice from the Santee Club.[90] "Col. Robert L. Montgomery feeds his wild ducks instead of shooting them," remarked Archibald Rutledge. Montgomery grew rice at Mansfield until 1943, when the war caused a local labor shortage.[91]

These are clear indications that rice planting long outlived its commercial viability in South Carolina as a form of waterfowl management. The career of Ludwig Beckman is a powerful case in point. Beckman sold his plantation to a member of the Santee Club in 1905, yet as an employee of the club, he continued to make a living planting rice for the next thirty-six years. Northern duck hunters initially took up rice planting because they viewed baiting as a legitimate mode of taking waterfowl and lowcountry plantations possessed large, experienced work forces that could produce bait relatively inexpensively. Rice cultivation continued on northern-owned plantations as long as wages remained low and baiting remained legal. Even if baiting had not been outlawed in 1935, growing rice "simply for the ducks" probably would not have survived much beyond World War II. Rising labor costs would have made the practice prohibitively expensive. Of course, the Santee-Cooper Project killed traditional rice culture on the Santee and the Cooper, where the discharge from the former drowned all of the fields, but elsewhere, it passed peacefully as the sportsmen who had resuscitated and sustained it during the baiting era gradually decided to end life support in favor of modern waterfowl-management techniques.

NOTES

1. For example, Ann Vileisis, Michael W. Giese, and Robert Pasquill, Jr., examine technical aspects of migratory-bird refuge development during the New Deal. See Vileisis, *Discovering the Unknown Landscape: A History of America's Wetlands* (1997; repr., Washington, D.C.: Island Press, 1999), p. 179; Giese, "A Federal Foundation for Wildlife Conservation: The Evolution of the National Wildlife Refuge System, 1920–1968" (Ph.D. diss., American University, 2008), pp. 154–159; Pasquill, *The Civilian Conservation Corps in Alabama, 1933–1942: A Great and Lasting Good* (Tuscaloosa: University of Alabama Press, 2008), pp. 207–213. Nancy Langston and Fredric L. Quivik treat complex resting- and breeding-ground restoration projects at refuges in Oregon and North Dakota, respectively, in the 1930s and 1940s. See Langston, *Where Land and Water Meet: A Western Landscape*

Transformed, Weyerhaeuser Environmental Book (Seattle: University of Washington Press, 2003), pp. 91–116; Quivik, "Engineering Nature: The Souris River and the Production of Migratory Waterfowl," *History and Technology* 25, no. 4 (Dec. 2009): pp. 307–323. Robert M. Wilson and Philip Garone investigate how irrigation and monoculture came to dominate waterfowl habitat on Oregon and California refuges after World War II. See Wilson, *Seeking Refuge: Birds and Landscapes of the Pacific Flyway*, Weyerhaeuser Environmental Book (Seattle: University of Washington Press, 2010), pp. 44–64, 79–94, 99–127; Garone, *The Fall and Rise of the Wetlands of California's Great Central Valley* (Berkeley: University of California Press, 2011), pp. 135–163. See also Wilson, "Directing the Flow: Migratory Waterfowl, Scale, and Mobility in Western North America," *Environmental History* 7, no. 2 (Apr. 2002): pp. 247–266.

2. Giese, "Federal Foundation for Wildlife Conservation," p. 152; Ira N. Gabrielson, "The Fish and Wildlife Service: A Summary of Recent Work," *Scientific Monthly*, Sept. 1947, p. 185. On the intersection of waterfowl ecology, wetland management, and national conservation policy in the 1920s and 1930s, see David L. Lendt, *Ding: The Life of Jay Norwood Darling* (Ames: Iowa State University Press, 1979), pp. 63–87; Jared Orsi, "From Horicon to Hamburgers and Back Again: Ecology, Ideology, and Wildfowl Management, 1917–1935," *Environmental History Review* 18, no. 4 (Winter 1994): pp. 19–40; Giese, "Federal Foundation for Wildlife Conservation," pp. 88–159. The foregoing literature was produced by historians. Also of value is the work of wildlife biologist Eric G. Bolen, the foremost student of the history of waterfowl management in the scientific community. See Bolen, "Waterfowl Management: Yesterday and Tomorrow," *Journal of Wildlife Management* 64, no. 2 (Apr. 2000): pp. 323–335. See also Guy A. Baldassarre and Bolen, *Waterfowl Ecology and Management*, 2nd ed. (Malabar, Fla.: Krieger Publishing Co., 2006), pp. 2–16, 470–507. For a more detailed international perspective, see John Paul Morton, "Duck Diplomacy: U.S.-Canadian Migratory Waterfowl Management, 1900–1961" (Ph.D. diss., University of Southern California, 1996), pp. 1–154; Kurkpatrick Dorsey, *The Dawn of Conservation Diplomacy: U.S.-Canadian Wildlife Protection Treaties in the Progressive Era*, Weyerhaeuser Environmental Book (Seattle: University of Washington Press, 1998), pp. 165–237. For a history of the concept of carrying capacity in wildlife biology, see Christian C. Young, "Defining the Range: The Development of Carrying Capacity in Management Practice," *Journal of the History of Biology* 31, no. 1 (Spring 1998): pp. 61–83.

3. John M. Anderson and Frank M. Kozlik, "Private Duck Clubs," in *Waterfowl Tomorrow*, ed. Joseph P. Linduska (Washington, D.C.: U.S. Department of the Interior, Bureau of Sport Fisheries and Wildlife, Fish and Wildlife Service, 1964), pp. 519, 523; J. Clark Salyer II and Francis G. Gillett, "Federal Refuges," in ibid., p. 505.

4. See Matthew Allen Lockhart, "From Rice Fields to Duck Marshes: Sport Hunters and Ecological Change on the South Carolina Coast, 1890–1940" (Ph.D. diss., University of South Carolina, in progress), chap. 3.

5. "A Duck Preserve," *Los Angeles Times* (Los Angeles, Calif.), July 27, 1896, p. 5.

6. Lockhart, "From Rice Fields to Duck Marshes," chap. 3.

7. See, for example, George C. Rogers, Jr., *The History of Georgetown County, South Carolina* (Columbia: University of South Carolina Press, 1970); Dennis T. Lawson, *No Heir to Take Its Place: The Story of Rice in Georgetown County, South Carolina* (Georgetown, S.C.: Rice Museum, 1972); James M. Clifton, "Twilight Comes to the Rice Kingdom: Postbellum Rice Culture on the South Atlantic Coast," *Georgia Historical Quarterly* 62, no. 2 (Summer 1978): pp. 146–154; Lawrence S. Rowland, "'Alone on the River': The Rise and Fall of the Savannah River Rice Plantations of St. Peter's Parish, South Carolina," *South Carolina Historical Magazine* 88, no. 3 (July 1987): pp. 121–150; Peter A. Coclanis, *The Shadow of a Dream: Economic Life and Death in the South Carolina Low Country, 1670–1920* (Oxford: Oxford University Press, 1989); James H. Tuten, *Lowcountry Time and Tide: The Fall of the South Carolina Rice Kingdom* (Columbia: University of South Carolina Press, 2010).

8. For a list of the charter members of the club as well as the active membership roll in 1900, see Henry H. Carter, *Early History of the Santee Club* ([Boston?]: privately printed, [1934?]), p. 4. For a complete member listing, see B. Brannan Reath II, *Santee Club—A Legend* (Philadelphia: Winchell Co., 1971), pp. 7, 9, 10–13. Carter and Reath were members of the club. Their anecdotal treatments contain numerous errors of fact and are most useful when consulted alongside other sources. For more on the donation of the Santee Club preserve to the Nature Conservancy, see Jack Leland, "Large Land Tract Given Conservancy," *Evening Post* (Charleston, S.C.), June 26, 1974, pp. A1–A2; Robert P. Stockton, "Conservancy Given 25,000-Acre Tract," *News and Courier* (Charleston, S.C.), June 27, 1974, p. A1; John Devlin, "Nature Conservancy Given 25,000-Acre Tract in South Carolina," *New York Times* (New York, N.Y.), June 27, 1974, p. 8; "Santee Coastal Preserve Established," Biosphere, *South Carolina Wildlife*, Sept.–Oct. 1974, p. 64; John Davis, "A Special Place," ibid., pp. 49–56; John V. Dennis, "Past and Present at Santee Coastal Reserve," *Nature Conservancy News*, Fall 1974, pp. 11–17.

9. "Dr. Emerson's New Club," *Baltimore American* (Baltimore, Md.), June 3, 1899, p. 4. On Emerson, see "Capt. Emerson Dies at His Valley Home," *Sun* (Baltimore, Md.), Jan. 24, 1931, pp. 18–19.

10. On Mackay, see *Dictionary of American Biography, Supplements 1–2: To 1940*, s.v. "Mackay, Clarence Hungerford," available online in the Gale Biography in Context database, BT2310008542, accessed Jan. 5, 2012. See also "Heir to Mackay Millions," *Sun*, July 28, 1902, p. 8; Earle J. Grellert, "The Present Head of the Mackay Enterprises," *State* (Columbia, S.C.), Aug. 10, 1902, p. 4; "Clarence Mackay Dies at Home Here after Long Illness," *New York Times*, Nov. 13, 1938, p. 1. On Jordan, see "Eben D. Jordan Died Yesterday," *Boston Daily Globe* (Boston, Mass.), Aug. 2, 1916, pp. 1–2; "Eben D. Jordan Dies at Home," *Boston Journal* (Boston, Mass.), Aug. 2, 1916, p. 1, 3; "Eben D. Jordan Dies at His Summer Home," *Boston Herald* (Boston, Mass.), Aug. 2, 1916, p. 1, 3; "Eben D. Jordan's Estate Totals $5,569,015," ibid., Feb. 10, 1917, p. 14; *Dictionary of American Biography*, s.v. "Jordan, Eben Dyer," available online in the Gale Biography in Context database, BT2310010106, accessed Jan. 5, 2012.

11. On Robert Jordan, see "Robert Jordan Dies in His Paris Home," *New York Times*, Nov. 4, 1932, p. 19. On the Hoyts, see Frank W. Norcross, *A History of the*

New York Swamp (New York: Chiswick Press, 1901), pp. 98–101; "Walter S. Hoyt Dies in Roosevelt Hospital," *New York Times*, July 15, 1920, p. 7; "Edward Hoyt Dies; Leather Pioneer," ibid., Nov. 29, 1925, p. E13; "George S. Hoyt Dies in Stamford Home," ibid., Oct. 15, 1931, p. 20.

12. On Frank M. Carnegie, see Mary R. Bullard, *Cumberland Island: A History*, Wormsloe Foundation Publications, no. 22 (Athens: University of Georgia Press, 2003), pp. 179–248, 331 (n. 54); obituary, *New York Times*, Feb. 23, 1917, p. 11. On McLean (and his sons, who are mentioned in the text below), see *Dictionary of American Biography*, s.v. "McLean, William Lippard," available online in the Gale Biography in Context database, BT2310007781, accessed Jan. 5, 2012. See also "W. L. M'Lean Dead; Noted Publisher," *New York Times*, July 31, 1931, p. 17; "The Bulletin Left to M'Lean's Sons," ibid., Aug. 6, 1931, p. 9; Andrea Knox, "The Bulletin, from Birth in 1847 . . . to Its Death Tomorrow: It Was Read," *Philadelphia Inquirer* (Philadelphia, Pa.), Jan. 28, 1982, p. A10. On Jay Cooke II (along with his son, who is referenced below in the text), see Ellis Paxson Oberholtzer, *Jay Cooke: Financier of the Civil War* (Philadelphia: George W. Jacobs and Co., Publishers, 1907), p. 2: 464; Lydia J. Ryall, *Sketches and Stories of the Lake Erie Islands*, Perry Centennial ed. (Norwalk, Ohio: American Publishers Co., 1913), pp. 305–320; "Jay Cooke Foresees the Birth of Great Industrial Cities," *Philadelphia Inquirer*, Aug. 6, 1903, p. 1, 6.

13. The Standard Oil officer referred to here was Herbert L. Pratt; for Penn Mutual Life Insurance, William A. Law; for Corn Exchange National Bank and Trust, Paul Thompson; for Pennsylvania Railroad, William W. Atterbury; for Fidelity-Philadelphia Trust, Marshall S. Morgan; for Chemical Bank and Trust, Percy H. Johnston; and for Lehigh Coal and Navigation, Samuel D. Warriner. On Pratt, see John N. Ingham, *Biographical Dictionary of American Business Leaders, N–U* (Westport, Conn.: Greenwood Press, 1983), s.v. "Pratt, Charles"; "Herbert Lee Pratt," *Successful American*, Apr. 1903, pp. 226–227. Law was a native South Carolinian and graduate of Wofford College in Spartanburg. On Law, see "William A. Law Promoted," *State*, June 7, 1922, p. 11; "Law Athletic Field Is Name of Wofford Site," *Spartanburg Herald* (Spartanburg, S.C.), June 3, 1930, p. 5; "W. A. Law Is Killed in Hunting Mishap," *New York Times*, Jan. 22, 1936, p. 1. On Thompson, see "Heads Philadelphia Bank; Paul Thompson New President of Corn Exchange National," ibid., Feb. 16, 1937, p. 31; "Paul Thompson, Veteran Banker; Chairman of Corn Exchange National Bank and Trust Co. of Philadelphia Dies," ibid., Dec. 14, 1942, p. 23. On Atterbury, see *Dictionary of American Biography, Supplements 1–2: To 1940*, s.v. "Atterbury, William Wallace," available online in the Gale Biography in Context database, BT2310018570, accessed Jan. 5, 2012. Atterbury appeared on the front cover of *Time* magazine's Feb. 20, 1933, issue and was profiled in an article in the Business and Finance department entitled "State and Stakeholders," pp. 41–42, 44–46. On Morgan, see "Heads Philadelphia Bank; M. S. Morgan Becomes President of Fidelity Trust Company," *New York Times*, Feb. 16, 1937, p. 39. On Johnston, see Frank Wilson Nye, *Knowledge Is Power: The Life Story of Percy Hampton Johnston, Banker* (New York: Random House, 1956); "With Bank 25 Years: P. H. Johnston of the Chemical to Celebrate Tomorrow," *New York Times*, Aug. 26, 1942, p. 31. On Warriner, see "General

Strengthening of Stocks; S. D. Warriner Elected President of Lehigh Coal & Navigation Company," *Philadelphia Inquirer*, June 5, 1912, p. 9; "S. D. Warriner, 75, Coal Leader, Dead," *New York Times*, Apr. 4, 1942, p. 13.

14. John Vavasour Noel, "Santee Gun Club Conserves Game," *News and Courier*, Feb. 28, 1932, p. B6.

15. Carter, *Early History of the Santee Club*, p. 4. On Grover Cleveland's last trip to the Santee Club, which took place fifteen months prior to his death, see Cleveland to L. A. Beckman, Mar. 21, 1907, in ibid., p. 5. See also "Cleveland's Hunt Brought to Close," *State*, Mar. 22, 1907, p. 1. "After several days' visit to the club house of the Santee Gun club [*sic*]," the article reads, "something over 200 ducks were bagged" by the president and his party "without half trying [*sic*]. Mr. Cleveland killed over half of these."

16. "Do You Know Your Lowcountry? Santee Gun Club," *News and Courier*, Nov. 15, 1937, p. 10; Leland, "Large Land Tract Given Conservancy"; Carter, *Early History of the Santee Club*, p. 9; Reath, *Santee Club—A Legend*, p. 29; T. S. Palmer, *Private Game Preserves and Their Future in the United States,* U.S. Department of Agriculture, Bureau of Biological Survey, Circular No. 72 (Washington, D.C.: Government Printing Office, 1910), p. 6.

17. Charles F. Kovacik and John J. Winberry, *South Carolina: The Making of a Landscape* (1987; repr., Columbia: University of South Carolina Press, 1989), p. 73; Sam B. Hilliard, "Antebellum Tidewater Rice Culture in South Carolina and Georgia," in *European Settlement and Development in North America: Essays on Geographical Change in Honour and Memory of Andrew Clark Hill*, ed. James R. Gibson (Toronto: University of Toronto Press, 1978), pp. 101–104.

18. Kovacik and Winberry, *South Carolina*, p. 73; Hilliard, "Antebellum Tidewater Rice Culture," pp. 103–104.

19. Joyce E. Chaplin, *An Anxious Pursuit: Agricultural Innovation and Modernity in the Lower South, 1730–1815* (Chapel Hill: Institute of Early American History and Culture by University of North Carolina Press, 1993), pp. 228–251; Judith A. Carney, *Black Rice: The African Origins of Rice Cultivation in the Americas* (Cambridge, Mass.: Harvard University Press, 2001), pp. 78–98; S. Max Edelson, *Plantation Enterprise in Colonial South Carolina* (Cambridge, Mass.: Harvard University Press, 2006), pp. 72–80, 98–113; Coclanis, *Shadow of a Dream*, pp. 82–98; Daniel C. Littlefield, *Rice and Slaves: Ethnicity and the Slave Trade in Colonial South Carolina*, Blacks in the New World (1981; repr., Urbana, Ill.: University of Illinois Press, 1991), pp. 74–114; Philip D. Morgan, *Slave Counterpoint: Black Culture in the Eighteenth-Century Chesapeake and Lowcountry* (Chapel Hill: Omohundro Institute of Early American History and Culture by University of North Carolina Press, 1998), pp. 39, 61.

20. Hilliard, "Antebellum Tidewater Rice Culture," pp. 105–109.

21. Ibid., pp. 100, 108.

22. Walter Edgar, *South Carolina: A History* (Columbia: University of South Carolina Press, 1998), pp. 267, 269; Amory Austin, *Rice: Its Cultivation, Production, and Distribution in the United States and Foreign Countries*, U.S. Department of Agriculture, Division of Statistics, Miscellaneous Series, Report No. 6 (Washington, D.C.:

Government Printing Office, 1893), p. 76; M. Richard DeVoe and Douglas S. Baughman, *South Carolina Coastal Wetland Impoundments: Ecological Characterization, Management, Status, and Use* (Charleston: South Carolina Sea Grant Consortium, 1987), 1: v, 9.

23. Tuten, *Lowcountry Time and Tide*, pp. 22–74; Charles F. Kovacik, "South Carolina Rice Coast Landscape Changes," in *Proceedings of the Tall Timbers Ecology and Management Conference, Number 16: February 22–24, 1979, Thomasville, Georgia* (Tallahassee, Fla.: Tall Timbers Research Station, 1982), pp. 54–55; Austin, *Rice*, p. 76.

24. Edward King, *The Southern States of North America* (London: Blackie and Son, 1875), p. 434 (first quotation); David Doar, *Rice and Rice Planting in the South Carolina Low Country* (Charleston: Charleston Museum, 1936), p. 8 (second quotation).

25. Archibald Rutledge, *Plantation Game Trails* (Boston: Houghton Mifflin Co., 1921), p. 250.

26. Chalmers S. Murray to "Mr. [Herbert R.] Sass, [Feb.?] 1934, fol. 14, box 24/101, Herbert Ravenel Sass Papers, South Carolina Historical Society ("second invasion" quotation); Murray, *This Our Land: The Story of the Agricultural Society of South Carolina* (Charleston: Carolina Art Association, 1949), p. 200.

27. Constitution quoted in Robert Ober v. Santee Club, *Case on Appeal, State of New York, Supreme Court, Appellate Division—First Department* (Cooperstown, N.Y.: Arthur H. Crist Co., [1910?]), p. 65.

28. Theodore D. Jervey [Jr.] to R. H. Lucas, May 7, 1900, Jervey Letter Book, 1898–1900, pp. 432–433, fol. 1, box 255, Theodore Dehon Jervey Papers, South Carolina Historical Society.

29. Doar, *Rice and Rice Planting*, p. 41.

30. Carter, *Early History of the Santee Club*, p. 11.

31. Ibid., pp. 11, 14.

32. Ibid., p. 15.

33. Santee Club Bag Records, vol. 1, 1901–1909, p. 99, South Carolina Historical Society (hereafter SCBR).

34. Carter, *Early History of the Santee Club*, pp. 14–15.

35. SCBR, vol. 1, pp. 135–136 (quotation on p. 135).

36. River Basin Studies Staff, Fish and Wildlife Service, Region 4, "A Report on the Lower Santee River, South Carolina to Determine the Effects of Diversion on Waterfowl Resources," July 1947, p. 11, Santee fol., box 447, Record Group 22, U.S. Fish and Wildlife Service, Division of River Basin Studies, General Project Files, 1946–1970, National Archives II, College Park, Md.; Louis S. Ehrich to Reid Whitford, May 17, 1894, quoted in "Appendix M: Improvement of Waccamaw and Lumber Rivers, North Carolina and South Carolina, and of Certain Rivers and Harbors in South Carolina," in *Annual Report of the Chief of Engineers, United States Army, to the Secretary of War for the Year 1894* (Washington, D.C.: Government Printing Office, 1894), pt. 6, p. 1088; "Rice Culture Declining," *News and Courier*, Aug. 7, 1907, p. 7; H. L. Oliver, "Will Abandon Rice Planting," ibid., Sept. 15, 1908, p. 3.

37. Ludwig A. Beckman Journals, 1919–1924 vol., pp. 120–122, private collection (hereafter LABJ); SCBR, vol. 1, p. 158; Carter, *Early History of the Santee Club*, p. 19.

38. "Pay Roll for Farm Work Done at the Santee Gun Club," June 16, 1906, fol. 1905–1929, Santee Club Records, Village Museum, McClellanville, S.C. On Alexander, see Maury Klein, *Edward Porter Alexander* (Athens: University of Georgia Press, 1971); Michael Golay, *To Gettysburg and Beyond: The Parallel Lives of Joshua Lawrence Chamberlain and Edward Porter Alexander* (New York: Crown Publishers, Inc., 1994). See also Alexander's "South Island Log," 1890–1910, microfilm copy in the Edward Porter Alexander Papers, Southern Historical Collection, Louis Round Wilson Special Collections Library, University of North Carolina, Chapel Hill, N.C. (hereafter SIL).

39. Although it varied from plantation to plantation over time depending on soil conditions and a host of other factors, two and a half bushels of seed rice to the acre was a fairly standard application. See, for instance, A Constant Reader [pseud.], "Summary of the Various Plans of Planting Rice, as Furnished in the Answers to Mr. Washington's Queries," *Southern Agriculturalist*, May 1829, pp. 193–197; [Solon Robinson], "Mr. Robinson's Tour—No. 18," *American Agriculturalist*, June 1850, p. 187; R. F. W. Allston, *Essay on Sea Coast Crops; Read before the Agricultural Association of the Planting States, on Occasion of the Annual Meeting, Held at Columbia, the Capital of South-Carolina, December 3d, 1853* (Charleston: A. E. Miller, 1854), p. 32; "Modes of Growing Rice in South Carolina," *Pacific Rural Press* (San Francisco, Calif.), July 29, p. 1, and Aug. 5, p. 1, 1871.

40. L. A. Beckman, letter to editor, *News and Courier*, Jan. 31, 1908, p. 6.

41. "The Santee Gun Club," *Evening Post*, Mar. 27, 1911, p. 6.

42. "McClellanville News," ibid., June 21, 1916, p. 7. W. Hampton Graham of Woodville Plantation and David Doar of Harrietta Plantation jointly planted the other twenty-five acres.

43. "Splendid Hunting along the Santee," *News and Courier*, Oct. 12, 1917, p. 10. This article also mentions that "they [the Santee Club] use over five hundred bushels of corn of their own raising for duck bait, besides buying on the outside. They will bait more heavily than ever before this coming season."

44. "Many Deer Are Being Flushed," *Evening Post*, Aug. 2, 1918, p. 11; LABJ, 1919–1924 vol., pp. 17–20; "Proposals Invited," *Evening Post*, Dec. 5–6, pp. 22, 12, and Dec. 8–10, pp. 14, 10, 18, 1919 (quotation).

45. Reath, *Santee Club—A Legend*, p. 61.

46. LABJ, 1939 vol., June 12.

47. Ibid., 1924 vol., Apr. 24. On the task system, see Philip D. Morgan, "Work and Culture: The Task System and the World of Lowcountry Blacks, 1700 to 1880," *William and Mary Quarterly*, 3rd ser., 39, no. 4 (Oct. 1982): pp. 563–599; Peter A. Coclanis, "How the Low Country Was Taken to Task: Slave-Labor Organization in Coastal South Carolina and Georgia," in *Slavery, Secession, and Southern History*, ed. Robert Louis Paquette and Louis A. Ferleger (Charlottesville: University of Virginia Press, 2000), pp. 59–78; Leslie A. Schwalm, *A Hard Fight for We: Women's*

Transition from Slavery to Freedom in South Carolina (Urbana, Ill.: University of Illinois Press, 1997).

48. L. A. Beckman, letter to editor, *News and Courier*, Feb. 17, 1919, p. 7.

49. Harry M. Walsh, *The Outlaw Gunner* (Cambridge, Md.: Tidewater Publishers, 1971), p. 30.

50. "Mr. Cleveland at Spesutia," Game Bag and Gun, *Forest and Stream*, Mar. 17, 1892, p. 248; "Mr. Cleveland at Spesutia," *New York Times*, Mar. 19, 1892, p. 4. See also "Grover Cleveland," *Sun*, Mar. 11, 1892, p. 4.

51. "Disheartened Duckers," ibid., Dec. 21, 1894, p. 6.

52. "Great Flocks of Ducks," ibid., Oct. 25, 1899, p. 6.

53. SIL, p. 26.

54. For more on Cleveland's time in South Carolina and his relationship with Alexander, see Lockhart, "From Rice Fields to Duck Marshes," chap. 5.

55. SIL, pp. 133–134; James Henry Rice, Jr., "Sport of the Rich Is Shooting Ducks," *News and Courier*, Dec. 18, 1904, p. 20 (quotation). See also Rice, "Hunts That Were Hunts," ibid., Dec. 28, 1910, p. 4.

56. Lockhart, "From Rice Fields to Duck Marshes," chap. 5. See also note 15.

57. Reath, *Santee Club—A Legend*, p. 61.

58. LABJ, 1928 vol., Jan. 20; Carter, *Early History of the Santee Club*, p. 19.

59. The weight of a bushel of rough rice is taken from William M. Lawton, *An Essay on Rice and Its Culture, Read before the Agricultural Congress, Convened at Selma, Alabama, December 5, 1871* (Charleston: Walker, Evans and Cogswell, Printers, 1871), p. 7.

60. Walsh, *Outlaw Gunner*, p. 29.

61. Carter, *Early History of the Santee Club*, p. 15.

62. SCBR, vol. 1, p. 157.

63. Ibid., vol. 3, 1914–1923, Jan. 31, 1923.

64. "Season on Ducks Opens Tomorrow," *News and Courier*, Nov. 15, 1931, p. A2; SCBR, vol. 5, 1930–1941, Dec. 15, 1931.

65. A. S. Jr. [Alexander Sprunt, Jr.], "Ducks and Ducks," Woods and Waters, *News and Courier*, Feb. 22, 1932, p. 4.

66. "Baiting, Now Banned, Led to Killing of More than 660,000 Ducks in '34," Aug. 9, 1935, available online in the U.S. Fish and Wildlife Service's Historic News Releases database, accessed Nov. 22, 2014, http://www.fws.gov/news/Historic/News-Releases/1935/19350809.pdf.

67. L. A. Beckman to J. N. Darling, Oct. 24, 1935, "Santee-Cooper, Gp-Z, S.C., 1930–1938" fol., box 135, Record Group 22, U.S. Fish and Wildlife Service, Bureau of Biological Survey, General Correspondence, 1890–1956, Reservations, National Archives II.

68. Stanley P. Young to L. A. Beckman, Oct. 29, 1935, ibid.

69. C. S. Murray, "Kinloch, Exclusive Santee Gun Club," *News and Courier*, July 12, 1931, p. B7; Alberta Morel Lachicotte, *Georgetown Rice Plantations*, rev. ed. (1961; repr., Columbia: State Printing Co., 1967), pp. 181–183; Rogers, *History of Georgetown County*, p. 493.

70. Both the South Carolina Historical Society and the Georgetown County Library Archives hold small collections of Kinloch-related materials. See Kinloch Gun Club Records, South Carolina Historical Society (hereafter KGCR); Kinloch Gun Club Photograph Collection, Georgetown County Library Archives, Georgetown, S.C. The Kinloch Club Records in the Eugene du Pont Papers at the Hagley Museum and Library in Wilmington, Delaware, is more comprehensive. See also Murray, "Kinloch, Exclusive Santee Gun Club"; "Where Rice Has Been Grown for More Than Century [*sic*]," *News and Courier*, July 12, 1936, p. C3; J. M. L., Jr. [John M. Lofton, Jr.], "Do You Know Your Charleston? Kinloch Plantation," ibid., Oct. 3, 1938, p. 10.

71. Lachicotte, *Georgetown Rice Plantations*, p. 178; Kinloch Gun Club Account Book, 1913–1917, pp. 110–115, fol. 1, box 25/148, KGCR; Kinloch Gun Club, Monthly Material Report, [1915?], fol. 7, box 25/147, ibid.

72. Wayne Darlington to Russel [*sic*] M. Doar, Mar. 27, 1919, Wayne Darlington to R. M. Doar, Apr. 28, 1919, Wayne Darlington to Russell [Doar], Oct. 13, 1919, fols. 10–12, box 25/147, KGCR; LABJ, 1933 vol., May 20; ibid., 1938 vol., Apr. 12; ibid., 1939 vol., May 22 and 24. On Darlington, see "School Board Sets New Limits for Districts; Squawk Expected (30 Years Ago in the *Statesman*)," Pioneer Days and Tales, *Idaho Statesman* (Boise, Idaho), Oct. 6, 1940, p. 6; "Two Mines in Challis Region Form Corporation," ibid., Nov. 28, 1927, p. 7; "Mackay Enjoys Big Mine Boom," ibid., May 11, 1926, p. 2; obituary, *New York Times*, Oct. 22, 1942, p. 21. On Chadwick, see "E. G. Chadwick, 63, Realty Ex-Leader," ibid., Mar. 24, 1945, p. 17. On Beach, see "W. N. Beach, Hunted Big Game in Alaska," ibid., May 6, 1955, p. 23.

73. See C. S. Murray, "Annandale, Secluded in Georgetown," *News and Courier*, May 31, 1931, p. B3; Murray, "Northern Hunters Control Rice Hope," ibid., June 14, 1931, p. B7; Murray, "The Wedge on the South Santee," ibid., Sept. 20, 1931, p. A7.

74. Murray, *This Our Land*, p. 201.

75. W. L. McAtee to Seth Gordon, Apr. 24, 1931, "Santee-Cooper, Gp-Z, S.C., 1930–1938" fol., box 135, Record Group 22, U.S. Fish and Wildlife Service, Bureau of Biological Survey, General Correspondence, 1890–1956, Reservations, National Archives II.

76. Doar, *Rice and Rice Planting*, p. 5; Workers of the Writers' Program of the Work Projects Administration in the State of South Carolina, comp., *South Carolina: The WPA Guide to the Palmetto State* (1941; repr., Columbia: University of South Carolina Press, 1988), p. 289.

77. Tuten, *Lowcountry Time and Tide*, p. 32.

78. Reath, *Santee Club—A Legend*, pp. 32–33; LABJ, 1938–1945 vol., p. 70.

79. Miles O. Hayes and Jacqueline Michel, *A Coast for All Seasons: A Naturalist's Guide to the Coast of South Carolina* (Columbia: Pandion Books, 2008), p. 164. For additional background on the Santee-Cooper Project, see T. Robert Hart, Jr., "The Santee-Cooper Landscape: Culture and Environment in the South Carolina Lowcountry" (Ph.D. diss., University of Alabama, 2004).

80. LABJ, 1938–1945 vol., p. 70.

81. Ibid., 1919–1924 vol., pp. 89–90.

82. Archibald Rutledge, *Home by the River* (Indianapolis: Bobbs-Merrill Co., 1941), pp. 16–17.

83. Archibald Rutledge, *Santee Paradise* (Indianapolis: Bobbs-Merrill Co., 1956), pp. 16, 122.

84. Rutledge, *Home by the River*, p. 17.

85. Lachicotte, *Georgetown Rice Plantations*, pp. 21–22.

86. "Rich Hunters in Georgetown," *News and Courier*, Nov. 25, 1909, p. 1.

87. Elizabeth W. Allston Pringle, *Chronicles of Chicora Wood* (New York: Charles Scribner's Sons, 1922), p. 17.

88. Suzanne Cameron Linder and Marta Leslie Thacker, *Historical Atlas of the Rice Plantations of Georgetown County and the Santee River* (Columbia: South Carolina Department of Archives and History for the Historic Ricefields Association, Inc., [2001?]), p. 53; M. A. Boyle to Kinloch Gun Club, May 9, 1919, fol. 12, box 25/147, KGCR; James Henry Rice, Jr., "Great Waccamaw Marsh Tract Changes Hands at Georgetown," *Evening Post*, May 20, 1905, p. 2 (quotation). On Baruch, see Bernard M. Baruch, *Baruch: My Own Story* (New York: Henry Holt, 1957); Baruch, *The Public Years* (New York: Holt, Rinehart and Winston, 1960); Jordan A. Schwartz, *The Speculator: Bernard M. Baruch in Washington, 1917–1965* (Chapel Hill: University of North Carolina Press, 1981); James Grant, *Bernard M. Baruch: The Adventures of a Wall Street Legend* (New York: Simon and Schuster, 1983).

89. Special Report Number 5, "Waterfowl Conditions in the Coastal Area from Virginia to Georgia," fol. "N. Hotchkiss December, 1934," box 2, Record Group 22, U.S. Fish and Wildlife Service, Records of the Branch of Wildlife Research, Research Reports, 1912–1951, National Archives II.

90. LABJ, 1937 vol., Dec. 15. On Montgomery, see obituary, *New York Times*, Jan. 24, 1949, p. 19.

91. Archibald Rutledge, "The Old Plantations Live Again," *Saturday Evening Post*, Jan. 15, 1944, p. 44 (quotation); Lachicotte, *Georgetown Rice Plantations*, p. 84.

Chapter 5

Knowledge of the Hunt

African American Guides in the South Carolina Lowcountry at the Turn of the Twentieth Century

Hayden R. Smith

During the winter of 1899–1900, timber baron Joseph Hilton and a childhood friend hunted ducks in abandoned rice fields outside of Savannah, Georgia. Hilton found navigating the maze-like landscape all but impossible. An absence of landmarks and an intricate grid of canals and adjoining banks disoriented him and made finding *terra firma* difficult. Fortunately, he and his companion had the services of Cain Bradwell, "tall, angular and black, and well known about McIntosh County [Ga.] for his skill as a duck hunter, and his extraordinary knowledge of the abandoned rice fields." Cain led Hilton and his friend through the old canals in a pole-propelled boat, moving carefully and quietly so as not to startle waterfowl. After locating a desired field, Cain directed the two men into position so that they would be prepared to shoot when the ducks took flight. Cain knew exactly where to place the hunters for optimal advantage. He also showed great skill in navigating the rice fields, where tall embankments bordering the canals and grasses "that hid everything but a narrow strip of sky" made identifying landmarks nearly impossible. By skillfully locating sloughs—conduits between fields and the irrigation canals—Cain "could paddle along a canal, making right angle turns into other canals, until he reached the one he wanted that led to a broken trunk." As Hinton recounted, "before we could see the trunk, but [when we could] feel the water making a swift pace, Cain would say 'hol' fas now we is comin' out in de big ribbuh,' then turn suddenly into the opening of the old trunk and we would be swept through."[1]

This article examines the guiding services that African Americans such as Bradwell performed in the South Carolina lowcountry during the early twentieth century. Exclusive hunting clubs and wealthy northern plantation owners depended upon black guides to lead hunters to prime locations. African Americans' ability as hunting guides stemmed from deep knowledge of the lowcountry landscape and waterfowl ecology. Such knowledge secured a place for blacks in the post-emancipation political economy and gave them an advantage over outsiders, who lacked comparable expertise and had no way of attaining it quickly or easily. Ironically, knowledge that planters had encouraged under slavery became a source of power and autonomy for African Americans after emancipation. Before the Civil War, planters had encouraged slaves to hunt to protect the rice crop from feeding birds and to supply themselves with food for personal consumption, a measure that reduced planters' obligations. Hunting and ecological knowledge passed down from generation to generation became valuable tools that provided African Americans with employment opportunities on northern-owned hunting plantations after the Civil War.

At the heart of this history is the importance of technology transfer as it relates to changes in the built environment.[2] As economic and social forces dramatically shifted in the post-emancipation South, lowcountry rice plantation owners transformed their property from a monocrop industry to a landscape of leisure. African Americans continued to provide the labor required to maintain the plantation infrastructure. Workers maintained roads, built fences, planted crops, and repaired embankments. Many African Americans who worked as guides had previously labored as rice cultivators or descended from people who had labored in that capacity. Former rice cultivators knew how to create environments that attracted waterfowl and how to navigate the maze-like canal system that separated the fields. People whom whites had once valued for their ability to cultivate a cash crop thus became valued in new roles. Guides knew the subtleties of migration and feeding patterns that made waterfowl difficult to hunt. Many guides relied upon longstanding traditions rooted in rural lowcountry communities. Whether hunting for subsistence or in accordance with masters' directives, guides and their ancestors had become intimately familiar with the lowcountry landscape. Ducks figured among species that slaves hunted most frequently.[3]

African American hunting in the lowcountry is as old as Carolina. As historian Scott E. Giltner has observed, "African-Americans became so linked with Southern hunting and fishing that the two became almost inseparable."[4] During the early phases of Anglo settlement, enslaved Africans navigated the Carolina frontier, performing tasks such as harvesting timber, herding cattle, and hunting for pelts. Of these tasks, cattle ranching fostered the most detailed knowledge of the lowcountry environment. Livestock ranching became a precursor to colonial South Carolina rice cultivation. As historian John S. Otto has explained, "drawing upon British and African antecedents, cattle-ranching proved the ideal industry for early Carolina—a colony with an abundance of land and cattle but a shortage of capital and labor."[5] Cattle ranching took place within the three ecosystems that later became habitats for rice: upland longleaf pine communities, small stream floodplains, and low-lying hardwood bottomlands.[6]

Slaves' knowledge of the lowcountry landscape derived from the routines of cattle ranching. Hogs and cattle foraged freely "at no cost whatever" in upland forests and savannas during the summer and fed in hardwood bottomlands and marshland canebrakes during the winter. Enslaved cattle-hands' duties included rounding up free ranging livestock at specific times of the year. Fearful of disease, European colonists rarely ventured into low-lying swamps, preferring instead to delegate care of foraging livestock to enslaved Africans. In carrying out such duties, slaves developed special-ized knowledge of lowcountry ecosystems. In 1708, one writer noted that slaves knew "the Swamps and Woods, most of them Cattle-hunters." While planters attempted to define boundaries between plantations and the wilder-ness, slaves served as the "middling" between two environments. Everyday exposure to the environment gave slaves opportunities to put the landscape to work for their own benefit. By actively herding animals for their masters and sometimes escaping into the wilderness for brief reprieves, early cattle-hands moved easily between the pineland savannahs and the cypress bottomlands.[7]

Lowcountry slaves depended on hunting for food. In this regard, they dif-fered from their masters, who viewed hunting as a form of leisure. By the middle of the eighteenth century, slaveowners realized that having laborers supplement their diets with wild game and fish dramatically reduced food costs. Enslaved Africans often hunted with guns and traps provided by their masters. As historian Phillip Morgan explains, "no group of slaves could match those of the Lowcountry for the amount of time spent fishing and hunting."[8] Morgan's observation reflects the fact that lowcountry plantations employed the task system. Lowcountry planters managed their laborers by assigning slaves specific tasks each day. The task system differed from the gang labor employed on short-staple cotton plantations, for it allowed slaves greater autonomy and opportunities to work for themselves. Gang labor required slaves to work in groups for long hours under the watchful eye of an overseer or driver. The task system allowed slaves to carry out their pre-scribed duties largely of their own accord. Once the enslaved completed their assigned tasks, they spent the remainder of the day working as they wished. Slaves devoted much of their personal time to hunting and fishing. Charles Ball, for example, walked up to ten miles while hunting "raccoons, opossums, and rabbits as provided for two to three meals a week."[9]

Extensive hunting and fishing made generations of African Americans keen observers of the natural world. Because they depended on hunting for food, they—in the words of anthropologist Claude Lévi-Strauss—acquired

"extreme familiarity with their biological environment . . . [showed] passionate attention . . . to it and [also developed] precise knowledge."[10] In venturing
into unsettled territory, slaves left the order of the planter's domain and
temporarily freed themselves from planters' watchful eyes. Hours spent
hunting, trapping, and fishing gave slaves detailed knowledge of animal
habits and nature. Linguistics professor Patricia Jones-Jackson has noted that
twentieth-century Sea Island residents, many of them descendants of slaves,
"old and young, are fully acquainted with the ways of the local wild animals.
Their partial dependency on these animals as a food source has caused them
to pay closer attention to the animals' personalities and habits than would a
hunter from another area."[11] Enslaved parents passed their knowledge onto
their children in hopes that they, too, would become adept at hunting wild
game. Hunting not only served as a means of procuring food but gave slaves
temporary relief from oppression, no matter how fleeting it may have been.

Hunting led to status on lowcountry plantations. Planters chose skilled
slaves to accompany them on hunts and perform labor ranging from guiding
to menial physical tasks such as serving as a "rower of the boat and driver
of the hounds."[12] According to historian Theodore Rosengarten, enslaved
African Americans "were an essential part of any expedition," for they provided "knowledge of the environment as well as labor."[13] Slaves who accompanied their masters on day or multi-day hunts relieved themselves from the
monotony and drudgery of plantation labor.

Lowcountry blacks had exceptional command of bird hunting and the natural environment. "Rice plantations attracted millions of ducks," noted rice
planter J. Motte Alston. Migrating waterfowl fed on rice and other flora and
found desirable cover in the fields. Alston explained that the birds consumed
rice in such large volumes that they became "fat and finely flavored." Well-
fed birds provided hunters with attractive pray and planters with succulent
fare.[14] People who labored day-in and day-out in these ecosystems knew better than anyone the environments that waterfowl favored and what attracted
the birds to specific areas.

THE LOWCOUNTRY RICE LANDSCAPE

Rice cultivation has a close association with the lowcountry environment.
From inland rice fields nestled in small-stream floodplains and cypress
swamps to tidal rice plantations sprawled across broad low-lying river floodplains, cultivated micro-environments became the basis of elaborate systems
of water control, monocrop agriculture, and a social order based on slave
labor. Inland rice formed the basis of the South Carolina colonial plantation complex and enabled planters' participation in the Atlantic economy.

It also fostered dependence on enslaved labor and dramatic alterations of the natural landscape. In addition, growing numbers of enslaved Africans led to a diversely acculturated landscape unique to the Southeastern Coastal Plain. During the eighteenth century, the lowcountry became home to one of the most sophisticated hydrological landscapes in world history.

Inland rice fields typically consisted of two earthen dams enclosing a low-lying area bordered by ridges. Slaves built up embankments with fill from adjoining drainage trenches. The dam on the higher elevation harnessed stream or spring fed water to form a reservoir, or "reserve," that watered the lower rice fields. A second dam kept water on the fields to irrigate rice plants and kill off competing vegetation. Located between these two earthen structures was a series of smaller embankments and ditches that channeled and drained water during the growing cycle.[15]

During the antebellum period, a second type of rice plantation became dominant. Tidal cultivation relied on different topographic and hydrologic settings. By using the natural cycle of the ocean tides, tidal plantations increased output and mitigated the risk to crops posed by freshets and storms. All along the Carolina coast, rivers and streams rise and fall with the ocean tides for distances varying between five and twenty-five miles inland. By using the tidal cycle, planters harnessed the energy of freshwater rivers and streams for irrigation and drainage. Tidal fields claimed river swamp for cultivation. Permanent embankments and surrounding interior ditches kept high water out of fields, retained floodwater, and allowed fields to be drained as needed. As on inland fields, trunks controlled water flows. Trunks used on tidal fields, however, differed by allowing inward and outward flows of water. Planters designed gates that covered both ends of the trunk. When fields needed flooding, slaves opened the exterior gate (closest to the river); the interior gate pivoted on a hinge so that water flowed into fields from the force of the tide. Once the tide changed direction, slaves closed both gates to prevent impounded water from leaving the fields. After the desired time elapsed for irrigating the fields, slaves raised the interior gate to allow water to flow out of the pivoted exterior gate, thus preventing resurgent tidal waters from flowing into the fields. Tidal rice fields were subdivided into smaller plots to efficiently control water flow. By building levies on a grid system, planters directed flows of water with precision. These embankments were called quarter divisions because they originally encompassed a quarter of an acre. Planters connected these divisions to a network of canals, ditches, and drains to properly irrigate the crop.[16]

Rice fields attracted migratory birds in large numbers. Planters especially feared the bobolink, a species commonly called the "ricebird" or "maybird." Bobolinks moved in large groups that devoured rice crops with devastating frequency. Rice planter J. Motte Alston observed they "come in myriads and in flocks so dense as to cast a shadow on the green and golden fields." Another observer wrote, "with regards to the *Rice-Birds*, it is almost incredible what devastation these little creatures will make." The bobolinks' reputation is underscored by the fact that its scientific name, *Dolichonyx oryzivorus*, means "rice eating." Modern research reveals that the birds' diet is 76 percent rice.[17]

Bobolinks passed through the lowcountry on their northward migration in late April and May and returned again between mid-August and late September. The birds' migratory pattern coincided with stages of rice sprouting. Birds fed on tender rice seedlings during their spring migration and on the "soft and milky" grain in the fall. Planters altered planting schedules to avoid having rice in the fields when bobolinks passed through. Colonial naturalist Mark Catesby recounted one unprepared Ashley River planter losing forty acres of his fall harvest to rice birds. The extent of the devastation left him wondering "whether what [the rice birds] left was worth the expense of gathering in." Planters generally planted rice earlier to synchronize flood stages with ricebird migrations.[18] Colonial accounts, however, speculate that ricebirds damaged between 30 and 40 percent of the rice crop. Slaves worked earnestly to ward off birds during the growing season. Laborers called "bird-minders" worked the fields with muskets and whips to scare the pests away. In 1750, a correspondent for the *Gentleman's Quarterly* noted that "2, 3, or more negroes are constantly kept traveling from the time the rice begins to ear, until it is full enough to cut, through every rice field, up to their knees and waits in water, continually hallowing and beating any sounding things to keep these birds from alighting there on."[19]

Enslaved bird-minders acquired great skill in hunting. Duncan Clinch Heyward noted, "on every rice plantation, there was some one Negro who was known as the plantation duck hunter, and that was his only work."[20] Heyward's great-grandfather, Nathanial, had a duck hunter named Matthias who roamed his Combahee River rice fields, skillfully eradicating flocks of mallards feeding on the rice stubble. Matthias stealthily padded through the rice canals—a maze of waterways bisecting the river floodplains that he memorized through years of work. He went to great lengths to remain silent and camouflaged. He used rope instead of commonplace chain to anchor his canoe along the embankments, for example, specifically to avoid notifying birds of his presence. His skills included the ability to detect when the ducks would take flight, which allowed him to anticipate the ideal time to take aim and shoot. "These notes of alarm always caught the ear of Matthias," according to Heyward, "for he knew the ducks had sentinels on the looks out." With a "keen and practiced eye over the field," Matthias would silently crawl to the precise location to take aim and wait for the alarm to sound. Listening to the birds and strategically positioning himself made Matthias a renowned bird hunter.[21]

African American bird hunters lived on rice plantations throughout the lowcountry. These people commanded respect and trust and thus enjoyed greater status than typical field hands. Planters allowed them to carry arms and to hunt instead of working in the fields. Bird hunters consumed many of the birds they killed and occasionally sold game to neighbors.[22] Planter families also enjoyed birds killed by slaves. "Delmonico," wrote J. Motte Alston, "with all his art, could not produce a more delicate dish for breakfast than one of ricebirds in September." He added that "two or three dozen [were] simply cooked in a frying pan—no lard, no butter, seasoned only with a little salt. Their fat makes a dish full of yellow gravy, which with some Carolina 'small rice' . . . affords a repast which would inspire one to write love verses though his nature be prosy in the extreme."[23]

POSTBELLUM CHANGES IN THE PLANTATION ECONOMY AND AFRICAN AMERICAN LABOR

The Civil War caused the demise of plantation agriculture in the lowcountry. Emancipation sparked a restructuring of economic and social activity from which planters never recovered. The two most successful cash crops of the antebellum era, rice and sea island cotton, did not disappear overnight. Instead, agricultural landscapes shifted over time as landowners failed to turn profits and compete with new technologies and as natural disasters took a toll. Rice planters faced increasing domestic competition from new producers in Louisiana, Arkansas, and Texas by the final two decades of the nineteenth century. A series of devastating storms between 1893 and 1911 caused extensive damage, a problem compounded by the long-term effects of inadequate maintenance. Put simply, planters found maintaining the elaborate water control systems needed for large-scale rice production impossible without slave labor. The storms of 1893–1911 sent many planters into a financial tailspin. Paying for repairs proved impossible; lost crops left their accounts depleted. By about 1900, the lowcountry rice industry had become economically inconsequential and all but ceased to exist by about 1920.[24]

Sea island cotton fared somewhat better, but not by much. As late as 1907, the prospects for the staple looked promising, but growing competition from domestic and foreign producers and hybridization soon sent conditions spiraling downward. The arrival of the boll weevil in 1917 sounded the death knell for the crop. In 1918, South Carolina growers produced 10,358 bales, the last commercially significant harvest. Thereafter, planters turned to truck farming and allowed thousands of acres to lie fallow, joining those previously left idle by the decline of rice production.[25]

With the decline of the staples that had historically sustained the lowcountry economy, new forms of activity developed. Logging operations and phosphate mining supplanted rice cultivation throughout much of the region, for example. Land values declined with the downturn in agricultural productivity. Tidal plantations once valued at $100 to $150 per improved acre fell to $20 to $30 an acre after the Civil War and by 1911 could "be had for a nominal sum."[26] Low land prices, combined with bountiful wildlife and the "Plantation Mystique," led wealthy sportsmen from northern cities to purchase decaying rice lands and turn them into winter retreats and hunting preserves.[27]

Changes in the plantation landscape led African Americans to assume new roles. Former slaves and their descendants used their knowledge of water hydrology, agricultural schedules, and processing rough rice for new purposes. With the decline of commercial rice production, African Americans began farming as tenants and took jobs with logging operations, as stevedores, and as fishermen.[28] Demand for African Americans guides thus competed with other forms of employment.

African Americans employed as hunting guides became an important fixture on rural retreats around the turn of the twentieth century. Depending on prevailing conditions, guides took property owners and guests to duck blinds to shoot waterfowl or drove deer from cover and out into the view of waiting hunters.[29] Although both varieties of hunting required specialized skills, many guides proved adept at both.

Knowledge of the rural landscape became valuable as sport hunting intensified. Demand for skilled guides increased dramatically in the late nineteenth and early twentieth centuries. African Americans who had grown up on rice plantations knew the landscape intimately and possessed extensive knowledge of animal behavior. Virginia Christian Beach has shown that William Garrett, a hunting guide at the Santee Gun Club, had been born on lands that the club acquired and either lived on them or within a seven-mile radius his entire life. Garrett's father, George, "was a trunk minder." He "taught the same skills to William, who also became a hunting guide." Depending on a property owner's needs, guides found themselves handling diverse responsibilities.[30]

Redevelopment of former plantations as hunting retreats created unprecedented employment opportunities for rural blacks. By the last decade of the nineteenth century, men who had made large fortunes in industry, finance, and commerce began accumulating large landholdings in the South. In the lowcountry, for example, J. P. Clyde, head of the Clyde Line Shipping Company, bought most of Hilton Head Island in 1889. Two years later, United States Senator J. Donald Cameron, of Pennsylvania, bought Coffin Point on St. Helena Island. Harry B. Hollins, a New York banker, bought Good Hope and White Hall Plantations in Jasper County. In 1893, a conglomerate of New York, New Jersey, and Pittsburgh businessmen acquired 42,000 acres in Jasper County to form the Okeetee Club (it later grew to 62,000 acres). In comparison, according to James Kilgo, antebellum

South Carolina plantations averaged between 2,000 to 3,000 acres apiece. Consolidation of former plantations as large retreats placed new value on local knowledge and labor.[31]

Duck hunting was the primary sport of choice for these new landowners. Ideal conditions in former rice fields and associated wetlands led ducks to feed and nest along tidal floodplains. Ironically, antebellum rice planters had not viewed duck hunting as a socially acceptable activity. According to Duncan Clinch Heyward, "the reason why the planters before the Civil War did not care for duck shooting was that these birds were then so plentiful in their fields, and hence so easily killed, that their killing was not considered good sport. It was looked upon more as 'pot-hunting.'"[32] Gentleman hunters of the postbellum era viewed matters differently. The challenge of shooting birds on the wing—while in flight—made ducks favored prey, and well-developed tastes for wild game added to hunters' enthusiasm. As duck populations declined dramatically during the last quarter of the nineteenth century, sportsmen from northern cities sought out new hunting grounds. Overhunting caused by the growing popularity of sport hunting and market hunting decimated wildlife populations across the North and Middle West. Men who could afford to travel made the coastal South favored territory. With its decaying rice plantations and fertile swamplands, the Carolina lowcountry rewarded hunters with bountiful opportunities.[33]

The rise of hunting as a primary form of activity increased lowcountry landowners' reliance on local knowledge. Though both white and black locals served as guides, the large population of African Americans who resided on or in close proximity to rice plantations made them an ideal source of labor. Hunt clubs and owners of private retreats hired African Americans for a wide range of jobs. Records kept by clubs and landowners reveal roles that echoed antebellum precedents: "laborer," "regular hand," "teamster," "gardener," "carpenter," and "plowman," for example. Guides held the most respected and trusted of all positions. For duck hunts, guides took hunters to carefully placed blinds, rowed boats through marshlands and along former rice canals, advised hunters on when and where to shoot waterfowl, and retrieved downed birds.[34]

Successful guides earned notoriety and respect. When guides became known for their skill and knowledge, hunting clubs benefited. Ensuring that members and guests had successful hunts figured among their priorities, after all. African American guides thus supplied a crucial piece of the equation that turned lowcountry lands into prime hunting domains. Despite the respect that many guides earned, white descriptions invariably showed paternalistic thinking. Elizabeth Allston Pringle, for example, described a guide at Chicora Wood Plantation as "a splendid boatman and was as much at home in the water as a duck . . . [who] made an easy living, at the same time satisfying

Figure 5.1   A. S. Sally wrote fondly of Isaac "spending his week days guiding for hunters and his Sundays guiding the spiritual welfare [of local African Americans]." From A. S. Sally, Jr., *The Happy Hunting Ground: Personal Experiences in the Low-Country of South Carolina* (Columbia: The State Co., 1926). *Source*: Reprinted with permission of The State Media Group, Columbia, S.C.

his love for the sport by taking strangers out ducking."[35] Nicknames such as "old Adam" and "old John" reflected the double standard of paternalism and affection that characterized employers' views of plantation guides. Reminiscing about hunting at Butler's Island Plantation on the Ashepoo River in Colleton County, A. S. Sally, Jr., noted that Bill White, a guide, once "threw out of an open rice field trunk after he had fired two barrels into great masses of ducks that had flocked into the trunk when the rest of the waters of the field were frozen over," which resulted in "eighteen mallards bagged at one shot." Guides with skill and knowledge became part of hunting lore, enshrined in epic stories told by white hunters in conversation and in print.[36]

Despite white respect of African American guides and their knowledge of lowcountry environments, race relations cast an unequal shadow upon labor and treatment. African Americans did not receive the same pay as white guides. The Oakland Club advertised "colored guides can be obtained at $1.00 per day. White guides are more expensive."[37] Whites viewed rural African Americans' work as "unskilled." In failing to appreciate the complexity of knowledge acquired over generations, whites overlooked African Americans' knowledge and expertise.[38] Charleston *News and Courier* columnist Chlotilde R. Martin hinted at these patterns in 1932 when she described the seasonal influx of northerners to the South Carolina lowcountry. As she commented, "they themselves brought back the old days with their benevolent and paternalistic attitude to the negroes they found living upon the lands which they purchased for hunting preserves."[39] Northern landowners did little to improve the lives of African Americans who supported their leisure and sporting pursuits.[40]

Guiding also discriminated by gender. Men exclusively guided clients and received payment for their services, while women performed domestic roles such as cleaning ducks and cooking for white hunters. Similar to gender patterns among the enslaved, men exclusively handled hunting and guiding. Antebellum planters did not assign women to work as bird minders or direct them to hunt for subsistence. While much literature is attributed to the critical roles that women played in rice cultivation knowledge, their roles were limited in hunting culture. Instead, women "dressed" or cleaned shot birds, cooking the waterfowl for landowners or clients. In an image of African American guides at Eldorado Plantation on the South Santee River, Charleston County, South Carolina, four men stand attentive posing for the photographer, while a single woman cook stands in the background.[41]

CONCLUSION

Viewing the lowcountry landscape in ways that reach beyond the limited perspective of white elites recovers uses of knowledge and power that

Figure 5.2 Candid and posed images of Eldorado Plantation hunting guides, South Santee River, Charleston County. Note the woman cook in the second image. From A. S. Sally, Jr., *The Happy Hunting Ground: Personal Experiences in the Low-Country of South Carolina* (Columbia: The State Co., 1926). *Source*: Reprinted with permission of The State Media Group, Columbia, S.C.

scholars have long overlooked. The unique forms of knowledge that African Americans possessed led them to see lowcountry landscapes differently than whites. Intimate familiarity with the spaces and environmental conditions of rice cultivation made them keen observers of elaborate built environments that attracted waterfowl and other wildlife in huge numbers.

Knowledge of hydrology, animal behavior, game habitats, and weather patterns made African Americans skilled guides. White hunters valued African Americans' knowledge even as they saw them as primitive, childlike creatures. Paternalism and racism combined with appreciation of blacks' skill in navigating old rice fields, locating feeding grounds, and positioning hunters in favorable locations. Ultimately, hunting produced patterns of black-white interaction similar to those that characterized social relations throughout the rural South. Despite close interaction and centuries-old ties, members of each group more often misread the intentions and abilities of the other than not.

The history of African American guiding in the lowcountry calls to mind the divergent views of plantation landscapes that the architectural historian Dell Upton identified in his seminal essay, "Black and White Landscapes in Eighteenth-Century Virginia." In exploring the ideological dimensions of plantation landscapes, Upton highlighted sharp contrasts between whites' and slaves' perspectives. Whites experienced the great plantations of the Virginia tidewater as "articulated processional landscape[s]" that placed the planter at the center of affairs. Hierarchical and carefully ordered, such landscapes sought to convey the planters' authority and the subordinate status of slaves, dependents, and whites of lesser social status. Visitors recognized not only planters' power as rooted in plantations but linkages to social and governmental institutions that showed the gentry to possess authority as a group. By contrast, slaves saw a landscape that centered on their quarters—their domestic space—and emphasized places that masters saw as inconsequential. Upton noted that slaves placed particular value on woods and fields, which offered them a measure of seclusion and secrecy, and pathways that connected them to quarters on other plantations "and unofficial ties with friends, relatives, spouses, and lovers."[42] Rebecca Ginsburg has recently extended Upton's analysis in examining fugitive slaves' views and use of plantation landscapes. Ginsburg identifies a "black landscape" as a "system of paths, places, and rhythms" that a "community" of African Americans "created as an alternative" to whites' perception of the same landscape, "by which they made sense of and imagined their surroundings."[43] Together, Upton and Ginsberg show dramatically different perceptions of the same places and spaces, how readings of landscape serve to normalize social orders, and how specialized knowledge provides opportunities to challenge such hierarchies.

Placing Upton's and Ginsberg's findings in a post-emancipation context shows that perceptual differences remain just as important in understanding differing forms of knowledge and power. African American guides navigated more freely and confidently through the intricate maze of canals and embankments that characterized postbellum rice landscapes than whites. Guides' knowledge of such landscapes derived from the experience of slavery and slaves' historical efforts to distance themselves from their masters. When hired to guide white hunters decades later, blacks used deep-seeded knowledge for economic benefit. The limitations of whites' knowledge became readily apparent whenever hunters ventured into the fields and swamps.

Although black hunting guides suffered poverty, racism, and the many other hardships that rural African Americans endured, post-emancipation guiding shows how specialized expertise allowed them to exercise authority amid severe inequality and maintain cultural and economic autonomy. Despite the unpredictable conditions associated with changes in landownership and economic flux, people who lived in proximity to ancestral lands maintained skillsets that provided ample benefit through services rendered to a new generation of white plantation owners. Despite the stereotypes of ignorant and primitive people, hunting guides responded nimbly to dramatic changes in land use and economic activity. By using over two hundred years of accumulated knowledge and suffering in rice fields, the new generation of plantation residents used labor and landscape to the best of their ability in providing for and giving the next generation a better life than they knew.

NOTES

1. Hilton's recollection of Cain Bradwell quoted in J. William Harris, *Deep Souths: Delta, Piedmont, and Sea Island Society in the Age of Segregation* (Baltimore: Johns Hopkins University Press, 2001), p. 139.

2. On African American technology transfer as it relates to New World rice cultivation, see Judith Carney, *Black Rice: The African Origins of Rice Cultivation in the Americas* (Cambridge: Harvard University Press, 2001); Judith A. Carney and Richard Nicholas Rosomoff, *In the Shadow of Slavery: Africa's Botanical Legacy in the Atlantic World* (Berkeley: University of California Press, 2009); Edda L. Fields-Black, *Deep Roots: Rice Farmers in West Africa and the African Diaspora* (Bloomington: Indiana University Press, 2008); Frederick C. Knight, *Working the Diaspora: The Impact of African Labor on the Anglo-American World, 1650–1850* (New York: New York University Press, 2010); Gwendolyn Midlo Hall, "Africa and Africans in the African Diaspora: The Uses of Relational Databases," *American Historical Review* 115, no. 1 (Feb. 2010): 136–150.

3. On African Americans and environmental history, see Dianne D. Glave and Mark Stoll, eds., *"To Love the Wind and the Rain": African Americans and Environmental History* (Pittsburgh: University of Pittsburgh Press, 2006); Ras Michel Brown, *African-Atlantic Cultures and South Carolina Lowcountry* (New York: Cambridge University Press, 2012); Carolyn Finney, *Black Faces, White Spaces: Reimagining the Relationship of African-Americans to the Great Outdoors* (Chapel Hill: University of North Carolina Press, 2014); Andrew W. Kahrl, *The Land that Was Ours: African American Beaches from Jim Crow to the Sunbelt South* (Cambridge: Harvard University Press, 2012).

4. Scott E. Giltner, *Hunting and Fishing in the New South: Black Labor and White Leisure After the Civil War* (Baltimore: Johns Hopkins University Press, 2008), p. 128.

5. John S. Otto, "The Origins of Cattle-Ranching in Colonial South Carolina, 1670–1715," *South Carolina Historical Magazine* 87, no. 2 (Apr. 1986): p. 122.

6. Peter H. Wood, "They Understand Their Business Well: West Africans in Early South Carolina," in *Grass Roots: African Origins of an American Art*, ed. Dale Rosengarten, Theodore Rosengarten, and Enid Schildkrout (Seattle: University of Washington Press, 2008), pp. 87–89; Stephen Pyne, *Vestal Fire: An Environmental History, Told through Fire, of Europe and Europe's Encounter with the World* (Seattle: University of Washington Press, 1997), pp. 466, 477 (quotation); John S. Otto, "Livestock-Raising in Early South Carolina, 1670–1700: Prelude to the Early Rice Plantation Economy," *Agricultural History* 61, no. 4 (autumn 1987): pp. 13–24; Mart A. Stewart, "'Whether Wast, Deodand, or Stray': Cattle, Culture, and the Environment in Early Georgia," *Agricultural History* 65, no. 3 (summer 1991): pp. 1–28; Louis Thibou letter, 20 Sept. 1683, Louis Thibou Papers, South Caroliniana Library, Columbia, S.C.; John Oldmixson, *The British Empire in America, Containing the History of Discovery, Settlement, Progress and State of the British Colonies on the Continent and the Islands of Americas*, 2nd ed. (2 vols.; London: J. Brotherton et. al., 1741), I: 521.

7. On the early colonial disease environment, see Hayden R. Smith, "Rich Swamps and Rice Grounds: The Specialization of Inland Rice Culture in the South Carolina Lowcountry, 1670–1861" (Ph.D. diss., University of Georgia, 2012), pp. 34–36; Otto, "Livestock-Raising in Early South Carolina," pp. 15–20; Clarence L. Ver Steeg, *Origins of a Southern Mosaic: Studies of Early Carolina and Georgia* (Athens: University of Georgia Press, 1975), p. 106 (quotation); S. Max Edelson, "The Nature of Slavery: Environmental Disorder and Slave Agency in Colonial South Carolina," in *Cultures and Identities in Colonial British America*, ed. Robert Olwell and Alan Tully (Baltimore: Johns Hopkins University Press, 2006), pp. 22, 24, 27.

8. James H. Tuten, *Lowcountry Time and Tide: The Fall of the South Carolina Rice Kingdom* (Columbia: University of South Carolina Press, 2010), p. 43; Phillip D. Morgan, *Slave Counterpoint: Black Culture in the Eighteenth-Century Chesapeake and Lowcountry* (Chapel Hill: Omohundro Institute of Early American History and Culture by the University of North Carolina Press, 1998), p. 138; Charles Joyner, *Down by the Riverside: A South Carolina Slave Community* (Urbana, Ill.: University of Illinois Press, 1984), p. 100.

9. Morgan, *Slave Counterpoint*, p. 138.

10. Lévi-Strauss quoted in Wood, "They Understand Their Business Well," p. 79. On technology and communally held knowledge transfer, see Robert Voeks and John Rashford, eds., *African Ethnobotany in the Americas* (New York: Springer, 2013); Judith Carney, "Rice, Slaves, and Landscapes of Cultural Memory," in *Places of Cultural Memory: African Reflections on the American Landscape*, ed. Bryan D. Joyner (Washington, D.C.: United States Department of Interior, 2003), pp. 45–61; Fields-Black, *Deep Roots*, pp. 187–193.

11. Patricia Jones-Jackson, *When Roots Die: Endangered Traditions on the Sea Islands* (Athens: University of Georgia Press, 1987), p. 17.

12. Rosengarten, "Introduction," in William Elliott, *William Elliott's Carolina Sports by Land and Water* (1867; reprint, Columbia: University of South Carolina Press, 1994), p. xiv.

13. Theodore Rosengarten, *Tombee: Portrait of a Cotton Planter* (New York: William Marrow, 1986), p. 128.

14. Arney R. Childs, ed., *Rice Planter and Sportsman: The Recollections of J. Motte Alston, 1821–1909* (1953; reprint, Columbia: University of South Carolina Press, 1999), p. 75.

15. Smith, "Rich Swamps and Rice Grounds," pp. 50–55.

16. On the evolution of tidal irrigation, see Mart A. Stewart, *"What Nature Suffers to Groe": Life, Labor, and Landscape on the Georgia Coast, 1680–1920* (Athens: University of Georgia Press, 1996), pp. 98–116; Sam B. Hilliard, "Antebellum Tidewater Rice Culture in South Carolina and Georgia," in *European Settlement and Development in North America: Essays on Geographical Change in Honour and Memory of Andrew Hill Clark*, ed. James R. Gibson (Toronto: University of Toronto Press, 1978), p. 97; Joyce E. Chaplin, *An Anxious Pursuit: Agricultural Innovation and Modernity in the Lower South, 1730–1815* (Chapel Hill: Institute of Early American History and Culture by University of North Carolina Press, 1993), chap. 7; Peter A. Coclanis, *The Shadow of a Dream: Economic Life and Death in the South Carolina Low Country, 1670–1920* (New York: Oxford University Press, 1989), pp. 66–68, 96–97. Earlier works include Duncan Clinch Heyward, *Seed From Madagascar* (1937; repr., Columbia: University of South Carolina Press, 1993); J. H. Easterby, ed., *The South Carolina Rice Plantation, as Revealed in the Papers of Robert F. W. Allston* (Chicago: University of Chicago Press, 1945); David Doar, *Rice and Rice Planting in the South Carolina Lowcountry* (Charleston: Charleston Museum, 1936); James M. Clifton, ed., *Life and Labor on Argyle Island: Letters and Documents of a Savannah River Rice Plantation, 1833–1867* (Savannah: Beehive Press, 1978); David Duncan Wallace, *The History of South Carolina* (4 vols.; New York: American Historical Society, 1934); Lewis Cecil Gray, *History of Agriculture in the Southern United States to 1860* (2 vols.; 1933; repr., Glouster, Mass.: Peter Smith, 1958).

17. Brooke Meanley, *Blackbirds and the Southern Rice Crop, Resource Publication 100* (Washington, D.C.: United States Department of the Interior, Fish and Wildlife Service, 1971), p. 39; Alexander Sprunt, Jr., and E. Burnham Chamberlain, *South Carolina Bird Life*, rev. ed. (Columbia: University of South Carolina Press, 1970), p. 489; Richard D. Porcher, *A Teacher's Field Guide to the Natural History of the Bluff Plantation Wildlife Sanctuary* (New Orleans: Kathleen O'Brien Foundation, 1985), p. 20; Alston quoted in Childs, ed., *Rice Planter and Sportsman*, p. 76; "Devastations by Rice Birds. Charles Town; S. Carolina, Oct. 15," *Gentleman's Magazine*, Jan. 1, 1751, p. 10.

18. Catesby quoted in "Of the Rice–Bird," *Gentleman's Magazine*, Jan. 1, 1751, p. 11. Ricebirds are also discussed in Stewart, *"What Nature Suffers to Groe,"* pp. 161–162; Richard D. Porcher, "Rice Culture in South Carolina: A Brief History, the Role of the Huguenots, and the Preservation of its Legacy," *Transactions of the Huguenot Society of South Carolina* 92 (1987): p. 7; Timothy Silver, *A New Face on the Countryside: Indians, Colonists, and Slaves in South Atlantic Forests, 1500–1800*

(New York: Cambridge University Press, 1990), p. 152; Francis Harper, ed., *The Travels of William Bartram: Naturalist's Edition* (1958; repr., Athens: University of Georgia Press, 1998), p. 188. Duncan Clinch Heyward noted that "never was any planting done between the middle of April and the last of May. The reason for not planting after the tenth of April was the coming of the May birds, which, during the month of May, were on their way northward, after wintering in the far South, and could always be depended upon to appear in our rice fields. They seemed to travel on a regular schedule, and were always on time." Heyward, *Seed From Madagascar*, p. 31.

19. Quote in "Devastations by Rice Birds," p. 10. See also Heyward, *Seed From Madagascar*, p. 32; Childs, ed., *Rice Planter and Sportsman*, p. 76; Doar, *Rice and Rice Planting*, p. 27.

20. Heyward, *Seed From Madagascar*, p. 124.

21. Ibid., pp. 124–127.

22. Stewart, *"What Nature Suffers to Groe,"* p. 174; Tuten, *Lowcountry Time and Tide*, p. 84; Joyner, *Down by the Riverside*, pp. 100–101.

23. Alston, *Rice Planter and Sportsman*, pp. 75–78 (quote on p. 77).

24. On the demise of the lowcountry rice industry, see Tuten, *Lowcountry Time and Tide*, chaps. 2 and 3; Coclanis, *Shadow of a Dream*, chap. 4.

25. Charles K. Kovacik and Robert E. Mason, "Changes in the South Carolina Sea Cotton Industry," *Southeastern Geographer* 25, no. 2 (Nov. 1985): pp. 96–98; Richard Dwight Porcher and Sarah Fick, *The Story of Sea Island Cotton* (Charleston: Wyrick and Co., 2005), pp. 329–330.

26. Gray, *Agriculture in the Southern United States*, p. 642; Quote in W. E. McLendon, G. A. Crabb, M. Earl Carr, and F. S. Welsh, "Soil Survey of Georgetown County, South Carolina," in *Field Operations of the Bureau of Soils, 1911* (Washington, D.C.: Bureau of Soils, 1911), p. 521.

27. Tuten, *Lowcountry Time and Tide*, p. 112.

28. Harris, *Deep Souths*, pp. 142–144.

29. Giltner, *Hunting and Fishing*, chap. 4.

30. Virginia Christian Beach, *Rice and Ducks: The Surprising Convergence that Saved the Carolina Lowcountry* (Charleston: Evening Post Books, 2014), p. 69.

31. Robert B. Cuthbert and Stephen G. Hoffius, eds., *Northern Money, Southern Land: The Lowcountry Plantation Sketches of Chlotilde R. Martin* (Columbia: University of South Carolina Press, 2009), pp. xvi–xxii.

32. Heyward, *Seed From Madagascar*, p. 123.

33. Nicholas W. Proctor, *Bathed in Blood: Hunting and Mastery in the Old South* (Charlottesville: University Press of Virginia, 2002), pp. 21–22; James A. Tober, *Who Owns the Wildlife?: The Political Economy of Conservation in Nineteenth-Century America* (Westport, Conn.: Greenwood Press, 1981), chap. 3; Daniel Justin Herman, *Hunting and the American Imagination* (Washington, D.C.: Smithsonian Institution Press, 2001), chap. 17.

34. Heyward, *Seed From Madagascar*, pp. 129–131.

35. Patience Pennington [Elizabeth Waties Allston Pringle], *A Woman Rice Planter* (New York: Macmillan Co., 1913), p. 172

36. Heyward, *Seed From Madagascar*, p. 117; A. S. Sally, *The Happy Hunting Ground: Personal Experiences in the Low-Country of South Carolina* (Columbia: The State Co., 1926), pp. 2, 15.

37. Quoted in Giltner, *Hunting and Fishing*, p. 131

38. Stewart, *"What Nature Suffers to Groe,"* p. 241

39. Martin quoted in Cuthbert and Hoffius, eds., *Northern Money*, p. 2.

40. On paternalism towards blacks at the turn of the twentieth century, see Mary E. Miller, *Baroness of Hobcaw: The Life of Belle W. Baruch* (Columbia: University of South Carolina Press, 2006), chap. 3.

41. Sally, *Happy Hunting Ground*, p. 38

42. Dell Upton, "White and Black Landscapes in Eighteenth-Century Virginia," in *Cabin, Quarter, Plantation: Architecture and Landscapes of North American Slavery*, ed. Clifton Ellis and Rebecca Ginsburg (New Haven: Yale University Press, 2010), pp. 122–139.

43. Rebecca Ginsburg, "Escaping Through a Black Landscape," in *Cabin, Quarter, Plantation*, pp. 51–66.

Chapter 6

A "Sporting Fraternity"

Northern Hunters and the Transformation of Southern Game Law in the Red Hills Region, 1880–1920

Julia Brock

Northerners who came to southern states to build or buy winter hunting estates did so at a particular historical moment. They moved to the South after the ouster of federal rule and during the hardening of stringent segregation laws, a time when the South's so-called "new men" were swearing allegiance to an order that purported to move away from the grip of agriculture and make way for business and industry.[1] They came southward at the peak of a third-party challenge to the seemingly solid Democratic Party and eventually saw that challenge dismantled and defeated. At first glance, northerners remained distant from these events. By all appearances, they built insular communities that lay outside the contours of southern life. Yet sportsmen could not have established their estates without assistance from white and black southerners. Close relationships with both groups proved crucial to the development of northern hunting colonies. African Americans supplied agricultural and domestic labor, served as hunting guides, and performed vital roles in supporting northerners' preferred forms of recreation. White southerners aided northerners' efforts to buy and lease land, secure labor, and manage their estates. By virtue of the relationships they developed with white and black southerners, northerners became embroiled in conflicts over land, wildlife, and social order.

In the Red Hills region of southwest Georgia and northern Florida, development of northern-owned estates coincided with fierce debates over land use, hunting, and emergent ideas about conservation. White farmers protested northerners' acquisition of the region's best lands and hunting practices, which limited the availability of game for subsistence. Northern hunters and their southern allies also supported changes in state game laws.

Debates were couched rhetorically in competing versions of masculinity. Many of the small farmers who contested the northern hunting colony were still reeling from the defeat of Populism and clinging to ideas about mastery borne by small producerism and property ownership. These small-holders rhetorically linked manhood to white, productive labor, an inde-pendence that could only be found as head of a household of dependents (that included women and children).[2] Northern hunters and their southern allies, on the other hand, claimed identity as sportsmen—gentlemen who practiced manly restraint, exercised skill in the field, and championed new ideas about conservation at the expense of smallholders and customary hunting practices.

Instead of remaining isolated from southern society, northern hunters benefited from the lines of class and race they encountered in the South. The alliance between northern and southern sportsmen was paramount in the former's ability to establish and grow a hunting colony in the Red Hills region. Southern sportsmen led the movement for game laws and finally the establishment of the Georgia Department of Game and Fish. Non-elite hunt-ers allied with sportsmen in formulating laws that curtailed the mobility of African Americans. But smallholders were less sanguine about the ways in which the laws protected elite sportsmen at the expense of their ability to trap, raise, and sell game. The divisions caused by new laws point to the fissures in class among white men in the Jim Crow South. Although controlling black mobility united white hunters to some extent, the battle over mastery of game played out along class lines in the statehouse and in the field.

Ultimately, opposition from small farmers did not derail the elite vision for a winter sporting colony. But, despite the seeming solidity of the politi-cal order and the welcome provided to northern men, the South still saw challenges to the new order and to the presence of northerners. This essay details those challenges in Georgia, where a northern hunting colony in the southwestern corner of the state had implications for statewide policy shifts and fueled a rhetoric of resistance. To gain insight into the impact of northern hunters on southern communities, this essay will examine Thomas County, in the northern part of the Red Hills region, and the role of one of its notable citizens, H. W. Hopkins, in championing the creation of a local northern hunting colony.

The Red Hills region encompasses Leon and Jefferson counties, Florida, and Thomas and Grady counties, Georgia. The landscape is characterized by rolling hills, lime sinks, longleaf pine stands and oak hammocks, wiregrass, and winding streams.[3] Historically, Tallahassee, Florida, and Thomasville, Georgia, have been the main centers of cultural life in the region. After the Civil War, Thomasville became a favored destination for northerners seeking escape from the chill of winter.[4] Over time, tourism and large sporting estates owned by wealthy outsiders shaped Thomasville's economic and social landscape. Northerners patronized small businesses, employed local citizens, and started a number of small enterprises. By the 1890s, they had established the area as a premier destination for bird hunting and hound and horse trials. Northern and Midwestern families supported local schools, churches, and hospitals and eventually established a small historical society.

Thomasville first began attracting travelers in the 1870s. The promise of curative pine forests brought sufferers of respiratory ailments to the town to convalesce. Unlike places such as Asheville, North Carolina, which remained popular with health-seekers for decades, tourism in the Red Hills faded by the 1890s. Connections forged early on provided the foundation for a winter hunting colony, however.[5] By the 1890s, wealthy industrialists from the North and Midwest owned land throughout the Red Hills. Inexpensive land prices facilitated development of large estates, and a ready supply of local workers gave northerners access to cheap labor. Land purchases and estate development continued well into the twentieth century. By the 1950s, northerners owned fifty hunting plantations in the Red Hills.[6] As late as 1976 these estates encompassed a total of 350,000 acres.[7]

During Thomasville's stint as a health resort, the surrounding area became known for a seemingly limitless supply of wild game, particularly quail. Hunting, a male-dominated pursuit in the nineteenth century, was popular among men of all races and classes in the South. Many men pursued game for family or individual sustenance, some hunted for the fur and meat markets, and others for leisure. As Nicolas Proctor has shown, hunting came to constitute an important part of southern manhood; the woods and field became the backdrop for testing and proving prowess; self-control, or the ability of the hunter to remain self-possessed in the excitement of the hunt, the feat of mind over physicality; and mastery, what Proctor calls a "multifaceted concept" that "represented control over other people, animals, nature, and even death."[8] These "distinctively southern" qualities of hunting were reserved for white men, who used hunting to reaffirm caste privilege and dominion over women and slaves. But, enslaved men also drew meaning from hunting; they were also passionate about hunting and wild game served to supplement their food supply. Some enslaved men served as mentors to white male adolescents learning woodsmanship. But, because African and African American men were severely circumscribed in their mobility and access to firearms, the ability to hunt and draw from the tropes of manhood that hunting offered whites was extremely limited.[9]

All white southern men claimed stake in the affirmation of manhood that hunting provided, and this bond leveled the class dimensions of shooting game by a good measure. Any white man could acquire a gun and a dog, and, at least until the later nineteenth century, hunting laws favored the open range and not the individual property owner. After the Civil War, a distinction grew between those who hunted for leisure and those who hunted for necessity. This divide was visible in hunting practice. More men with means began to consider themselves "sportsmen," a specific type of hunter. These men were, thanks to advocates such as George Bird Grinnell, taking part in the creation of a national language forming around themes of sport and conservation. Grinnell's publication, *Forest and Stream*, among others like it, was central in propagating this language to a wide circulation of readers, North and South. The magazine codified the sportsman's ethos in articles that dwelt on the manliness of hunting and fishing and the values of sporting— sport as purely a pursuit of leisure (as opposed to those who hunted for money or subsistence), as a skill of marksmanship, and as an example of fair play. Baiting fields and poisoning streams, both still common in the South, stood at odds with the code of the true sportsman, who allowed game a fighting chance. An identity only available to elite, white men who had leisure time and capital for guns and dogs, the moniker of sportsman was reserved for a certain class, but described men from across the country. The subscribers of magazines like *Field and Stream* considered themselves a kind of fraternity and, in the late nineteenth century, shared a bond not limited by old war wounds and sectional division. A masculinity bound in elite sporting practice increasingly connected men across sectional lines.

In the 1870s, thanks to *Field and Stream* and other publications, a growing, cross-regional movement of sportsmen took up the cause of conservation. H. W. Hopkins, the mayor of Thomasville, participated in this movement, as did northerners who owned land in the Red Hills. Hopkins exemplified the hunter-as-sportsman ideal. One writer from a national hunting magazine declared emphatically that Hopkins was a sportsman "I wish all sportsmen could know." It is no small coincidence that he did much to attract northern hunters to Thomasville and helped to build the winter hunting colony.[10] In promoting the region, he followed his uncle's lead. Dr. T. S. Hopkins had promoted the region as a destination for tuberculosis patients, largely on the basis of his belief in the healthful benefits of pine resin.[11] The younger Hopkins favored shooting and dogs and invested time and money into these pursuits. In 1882, the Atlanta *Weekly-Constitution* reported that Hopkins had a large kennel under construction in downtown Thomasville—one local called it the "dog hotel." The "hotel" had room for one hundred dogs, a kitchen, a trainer's house, and an exercise ground. Hopkins used it for breeding and training setters, pointers, and hounds. The newspaper called Hopkins an "authority" on dogs and hunting and noted that he had introduced beagles to hunters in the area. His reputation for breeding and training the best dogs was already cemented in the early 1880s among northern hunters in Thomasville;

Figure 6.1   H. W. Hopkins (left) and A. H. Mason after a Red Hills quail hunt, 1917.
Hopkins was an avid bird hunter and dog breeder and, through his efforts to help build
a northern hunting colony, formed lifelong friendships with northern hunters such as
Mason. *Source*: Courtesy Thomas County Historical Society, Thomasville, Ga.

a local told the reporter that, "Hopkins's dogs work like clocks, and no yankee [*sic*] ever shoots over one without wanting to buy him."[12]

Hopkins also kept up with the latest in the sporting press and actively
engaged with other hunters through the pages of national periodicals. He sent,
for example, the wing, tail, head, and foot of a bird to the editors of *Forest
and Stream* for identification (their response: "The bird is a king rail, or fresh
water marsh hen").[13] Hopkins connected to a national network of sportsmen through journals such as *Forest and Stream*, and he cemented relationships with northern hunters who came to Thomasville as tourists by hosting
foxhunts and shooting parties and by loaning and selling his hunting dogs.
Hopkins' enthusiasm for sport paved the way for a winter hunting colony.

The kind of hunting that Hopkins promoted required capital. His dogs cost
between $100 and $500, which put them out of the reach of most local

hunters.[14] As Thomasville gained a reputation for abundant game, the area increasingly drew northerners who sought the quarry of its fields and woods instead of a healthful resort town. These sportsmen transformed hunting into a spectacle, replete with the best dogs, guns, wagons, and an entourage of other hunters, wagon drivers, and dog trainers. Northerners found a southern counterpart in Hopkins and, like him, participated in the national culture of sporting that grew in the nineteenth century. They too subscribed to hunting periodicals and sought the best guns and dogs, and they increasingly pursued game in places outside of the northeast and, eventually, outside of the United States. Serious sportsmen had been coming South since antebellum times as their own lands were depleted of game.[15] Northern and southern hunters had shared the field before they formed networks in Thomasville and a code of sportsmanship that came of age after the Civil War.[16]

But these northern hunters were products of the Gilded Age. Many of them made fortunes from the new industries that dominated American business enterprise in the late nineteenth century: oil, railroads, and steel. For these men, hunting was more than just a pastime; it was charged with an ethos of wealth and domination. Historians have argued that hunting by wealthy sportsmen was directly tied to the age of empire in the late nineteenth century in Britain. Hunting, for example, often preceded or went hand-in-hand with territorial domination; the mastery of another territory and its fauna was congruent with the control of its people and institutions. In the United States, sportsmen "served empire in another way," as Daniel Justin Herman argues, by continuing to associate hunting and white American manhood and casting it in the light of late nineteenth-century ideas about scientific organization and racial hierarchy.[17] Theodore Roosevelt perhaps best personifies these themes; he idealized the western hunting adventure and later, traveling for the Smithsonian, the big game safari. Other wealthy hunters followed his lead onto western lands and eventually to faraway places to pursue sport. These sportsmen, like those who built the winter hunting colony in the Red Hills region, took part in a social drama that reaffirmed the power of wealth and racial hierarchy. Their capital bought adventure, a chance to prove manliness, and, by using a cadre of subordinates as helpmates, the opportunity to be a paternal master of the hunt. These men traversed the country in plush, private Pullman cars (they might have even owned the railroad itself) in search of prey and adventure. When they first came to southwest Georgia, they found not only abundant game, but a place where political and economic systems had been shaken, where farmers were impoverished, and where there existed a labor supply that was large and cheap. The South offered opportunities for wealthy sportsmen with visions for a genteel but rugged life. There already existed a hunting tradition that privileged white manhood; sportsmen found in the South and in its local elite, men like Hopkins, the perfect backdrop with which to create an idealized leisure community with a stable social hierarchy that blended a mythologized past with the modern.

Like hunting colonies in South Carolina, northerners bought the lands and homes of people who were once scions in the area's planter class. To acquire or increase landholdings, sportsmen relied upon their friendship with the local grandee, H. W. Hopkins. Hopkins combined extensive knowledge of local land and people with a business savvy that resulted in the formation of the winter hunting colony. He was an influential man who had the ability both to secure land at good prices and to inform northerners of local and state laws that would affect their property and hunting customs. By 1879, Hopkins had established a real estate company that formalized his role as a local agent. With his assistance, wealthy sportsmen acquired private hunting preserves and also leased shooting land into the 1930s.

Most of Hopkins' business began from social connections made in Thomasville and continued through word of mouth. In many cases, interested investors sought him ought rather than vice versa. In 1901, for example, D. L. Shepard of St. Paul, Minnesota, wrote Hopkins about a potential buyer in "an old and esteemed friend Mr. Marvin Hughitt Paes of C. + N.W.R.R. [Chicago and Northwestern Railroad]." Shepard "told him about Thomasville and the Keifer place and he was impressed very favorably. . . .. He is decidedly such a man as you would like to add to your Northern Colony."[18] If interested buyers secured an introduction to Hopkins and made a trip to the Red Hills region to survey available property, Hopkins put himself at their disposal. Charles S. Hebard of New Jersey, owner of Ty-Ty plantation near Thomasville, wrote Hopkins in 1903 thanking him for his careful attention to the interests of buyer J. H. Wade of New York. Wade, wrote Hebard, "seems pleased with [the property] and with the way you treated him—he is a very fine man and will be a great acquisition to the place."[19] The correspondence suggests that before a land sale took place, buyers such as Paes and Wade had to be satisfied with Thomasville and what it had to offer, and to win Hopkins's tacit approval.

Like the hunting enclaves that formed in the lowcountry, northerners reveled in the mythology of the Old South and exploited their opportunity to own a piece of it. Sketching the genealogy of several purchases illustrates this point. Dr. J. T. Metcalfe, a doctor and native New Yorker, spent his winters in Thomasville and, as historian William Rogers notes, "was a tireless promoter of the area's advantages."[20] Metcalfe's first land purchases in 1883 were in the southeastern portion of Thomas County; he bought the old plantation lands of James L. Seward, a prominent state congressman.[21] Though he sold these lands to David McCartney of Wisconsin in 1886, Metcalfe once more purchased land in 1887, the 1,600-acre Cedar Grove Plantation, from the Blackshears, one of the oldest and largest planter families in the area. Metcalfe's purchase included the original plantation home, which he renamed Susina, for his wife Susan. He did not remain long at Susina—he sold the property in 1891 to A. H. Mason, the heir to a shoe blacking business in Philadelphia—but Metcalfe had garnered enough influence in the area to become the namesake of a railroad stop created in 1889, Metcalf (the town later dropped its final 'e').[22]

Another early buyer was John W. Masury, a wealthy paint manufacturer from New York who had also built a hotel in Thomasville to cater to tourists. In 1887 he purchased a 1,500-acre property that he named Cleveland Park, where he often hosted picnics and parties for wealthy northern and southern whites.[23] The land had once belonged to another branch of the Blackshear family.[24] In 1889, S. R. Van Duzer, also from New York and a "millionaire," according to the local press, bought a 1,300-acre plantation, Greenwood, owned by the Jones family, another prominent planting family.[25]

The Hanna family, wealthy oil refiners from Cleveland (who sold out to Standard Oil in 1876), and their associates (partners in business and family

Figure 6.2 Greenwood Plantation, Thomas County, Georgia, circa 1930. English architect John Wind designed the Greek Revival home in the late 1830s for Thomas and Lavinia Jones, a prominent planting family in Thomas County. The Jones family sold the home to a New Yorker, S. R. Van Duzer, in the late nineteenth century. *Source:* Courtesy Thomas County Historical Society, Thomasville, Ga.

friends) came to dominate landholding in the Red Hills region. "By 1976," notes geographer William Brueckheimer, "the descendants of the Hannas, Hanna Company executives, and Cleveland friends owned forty-one plantations containing over 150,000 acres."[26] Salome Hanna, the sister of Howard Melville (H. M.) and Mark Hanna, made early purchases. She and her husband, J. Wyman Jones (who developed Glen Arven Country Club in Thomasville), bought a plantation in 1891 that they named Elsoma. The same year, Salome's son by her first marriage, Charles M. Chapin, purchased Melrose Plantation from a prominent local family. He later acquired Elsoma for himself. H. M. Hanna, a Standard Oil director who also ran the M.A. Hanna Company (a coal, iron ore mining, and shipping conglomerate) with his brother Mark, purchased Pebble Hill Plantation, an antebellum estate once owned by Thomas Jefferson Johnson, a founder of Thomas County.[27] During the 1880s and 1890s, then, lands in Thomas County and the Red Hills transferred from southern to northern ownership. Local families such as the Jones, Blackshears, and Johnsons—who had built their fortunes on cotton and slaves—sold to northern families whose wealth came from the booming industrial economy.

This lineage of former owners appealed to northerners captivated by the romance of the Old South. For sportsmen, antebellum homes symbolized a bygone aristocracy and fast-disappearing gentility. According to a former director of the Georgia Historic Sites Survey, the classical revival homes in Thomas County that became winter hunting estates "fit the dream ideal of the antebellum South better than those from any other part of Georgia."[28] Greenwood, the Van Duzer estate, later owned by the Whitney family, is perhaps the most famous. With its massive ionic columns, a two-story portico, and a hand-carved pediment, it stands as a temple to the agrarian social order. Many of the homes on northern hunting estates—Susina (Metcalfe's home until he sold it to the Mason family); Pebble Hill (owned by H. M. Hanna); Elsoma and Melrose—were antebellum in origin. Though they would install modern amenities, the northern owners largely left the facades of the homes unchanged (though a few of the homes, such as original house at Pebble Hill, later burned).[29] Now the resident gentlemen of these country estates, northern hunters were kings of leisure, not cotton.

Northerners purchased contiguous lands in order to expand their shooting domains. J. H. Wade provides a good example of the process. In 1904, Wade wrote to Hopkins agreeing to purchase the "Girtman place," a farm next to his Mill Pond lands.[30] In 1907, he purchased another parcel of contiguous land from a Miss McCartney of Green Bay, Wisconsin.[31] Three years later, he acquired two parcels owned by the McIntyres (known as the Futch lands), a prominent local family.[32] In 1916, Wade again wrote to Hopkins wishing to enlarge his holdings: "I would like buy the South ½ of lot 91 owned by

Mrs. Lillie if she will sell it at $15 per acre. This would connect my Futch land with the Hammond place I recently bought. Please see what you can do."[33] Lula Mae Hamilton, the Wade family governess, informed her mother that Wade loved "to buy the land and then go through laying out roads where he sees fit." She also noted that not all small farmers were willing to sell. Although Wade had successfully bought land from a few African American families and "let them live on" to farm shares, "There is one little place near here that two darkies own and won't give up some beautiful woods too."[34] Hamilton's offhanded slur belies the empowered stance of the smallholders who refused to sell out to the wealthy sportsman.

Elite hunting customs and business relationships cemented friendships between Hopkins and the northern sportsmen. Hopkins joined northerners on their own hunting grounds, on fishing expeditions in Florida, and sometimes even visited them in their home states. He also maintained hunting camps in the Red Hills region where he and northerners spent time hunting, eating and drinking, and telling tall tales. "Judge," as he was known affectionately, created lifelong friendships with men who served as the backbone of the winter colony.

Largely through Hopkins's efforts, consolidation of lands in the Red Hills proceeded swiftly. Clifton Paisley notes that by 1950, northern owners together held 109,700 acres in Leon County, a consolidation that reduced available agricultural land by eighty percent.[35] Because landholdings grew so large, only wealthy northern hunters could afford them when they went up for sale. In 1915, Hopkins conceded to northerner Edward Crozer that a "property like yours is beyond the average villager for a home at anything like it's value."[36]

Land consolidation angered locals. Not all farmers in Thomas County and the Red Hills region wished to sell out to wealthy northerners, and voices of dissent peppered local newspapers. In 1904, efforts to form a new county from parts of Thomas and Decatur Counties, for example, provided a vehicle for airing grievances against the Yankees.

Cairo was an emerging market town and railroad stop that served as a trading center for farmers in western Thomas County and eastern Decatur County. A former Populist stronghold, it retained a sizeable number of third-party sympathizers.[37] Logistical considerations sparked the initial push for a new county. As the editor of the *Cairo Messenger* explained, a new county "would be a great convenience for the people in this neighborhood, as this is another instance where the people have to go from 15 to 25 miles to reach the county site."[38] Traveling the fourteen miles to Thomasville or the twenty-two miles to Bainbridge (county seat of Decatur), the editor argued, was inconvenient and costly for farmers. Creating a new county would allow for new, more accessible municipal buildings and, at least in theory, would spur Cairo's growth.

A cadre of locals opposed the new county and its supporters. The editors of the *Thomasville Times-Enterprise and South Georgia Progress* led the charge. They used their columns to attack the movement and raise concerns about the dangers of breaking up two large counties. They argued that the measure would decrease revenue and thus lead to tax increases in Thomas County. They questioned the need for a new county seat and wondered if support for the initiative was a power grab by would-be politicians.[39] In more emotional terms, the faction also raised the issue of race and politics. At a public debate held in Thomasville in 1905, for example, a resolution created by opponents of the new county argued that because blacks made up a majority of the citizenry in Thomas County, a new county "would subordinate again their former associates and neighbors" to "this overwhelming mass of ignorance and idleness." Appealing to the new county supporters, the resolution entreated them to "have a human regard for the safety and well-being of their neighbors, who were their comrades in the long and bitter struggle [during Reconstruction] for white supremacy in Thomas County." Piggybacking on fears of whites becoming a racial minority, the anti-county movement referenced the presidential election of 1896, when Thomas had become the only county to vote a majority for McKinley, the Republican candidate. When H. W. Hopkins came to the defense of the new county with arguments of popular sovereignty, the editors in Thomasville accused him of "endeavoring to bring about a coalition of affairs by which Thomas county [*sic*] might become black Republican." Though 'black Republican' was a common epithet in the one-party system of southern politics, attacking Hopkins—whom all knew was an ally of the northern sportsmen, including the Hanna family, who had invited McKinley to Thomasville to meet with southern Republicans in 1896—was symbolic. Questioning Hopkins' appeal to republican principles, the editors complained of his "sophomoric . . . repetition of trite catch phrases 'vital principles of republics, essence of Democracy freedom and independence.'" They asked Hopkins, "Do you want to square your actions by a definition? Are you willing for white and black to vote? Did Webster know about the color line?"[40] Raising the specter of black political autonomy and subtly connecting it to wealthy northerners' influence on county politics, opponents of the new county relied upon bravado and fear to rail against the movement and its supporters.

Supporters of the new county counterattacked, going beyond arguments for convenience to pit the new county and its prospective population of small farmers against the landed interests in Thomasville. The attacks went to the heart of Thomas County's reputation as a hunting destination and its seasonal northern population. Countering the claim that a new county would raise taxes in Thomas, Grady County supporters wondered why they should "any longer help to pay taxes to keep up Thomasville and to work the Thomasville roads so their 'distinguished winter visitors' can air themselves luxuriously around in rubber tire carriages and four horse tallyho's?" The editor continued that, "If Thomasville has let her winter birds roost, and set, on all the land around there, driving out home people from their little farms . . . who is responsible for it?" Proponents of the new county pit the "foreign and privileged millionaire class" who had "gobble[d] up . . . lands" against the "home people" who were shut out of the "rich soil . . . which surrounded" the town.[41] In the rhetoric of heated argument, Grady County supporters cast themselves as the heirs of a Jeffersonian republic of small farmers and the "distinguished winter visitors" as a land-hungry elite who earned their wealth from "favored trusts."

Residents of Leon County, Florida, echoed concerns about northern land consolidation. In the Tallahassee *Weekly True Democrat*, one writer compared the game preserves of the Red Hills to those of England and noted that both had forced out small farmers. A 1914 editorial in the paper argued that,

> As much as the *True Democrat* appreciates the good judgment of wealthy men buying up large landed interest in Leon County for game preserves, it prevents the prosperity we are so anxious to see. Small farms are the true source of dependence, and the policy that prevents an increase of population is wrong and damaging.[42]

The editor also expressed a desire to see "the adoption of some plan whereby the large landed interests of Leon County could be converted into small, profitable farms."[43] By 1920, this vision had gone unfulfilled, prompting the editor to lodge another complaint: "Leon County is suffering much because large landlords are not bringing their immense acreage into production of needed crops."[44]

The editor pointed to a growing trend among northern landowners to reduce crop cultivation in favor of game conservation. This change was largely due to the decrease in the quail population, which became acute in the second decade of the twentieth century. Concerned about the lack of game, a group of hunters (including Charles Chapin, L. S. Thompson, owner of Sunny Hill Plantation, and Arthur B. Lapsley, owner of Meridian Plantation) hired the services of naturalist Herbert L. Stoddard to study the quail population and offer remedies to its decline.[45] Stoddard published his results in *The Bob-white Quail*, which became the preeminent guidebook for protection of the bird. Stoddard's management techniques dismissed commercial agriculture, particularly cotton cultivation strategies. Large-scale agriculture depleted the soil, deteriorating the food supply for quail and leaving them with no cover for a habitat. Instead, less-intensive "patch-style" agriculture (small plots of cultivated land separated by brush or tree stands) was the best environment for the bird to thrive.[46] This directive was an incentive for northern hunters to maintain the sharecropping system, which employed patch-style farming, but to allow for less intensive agricultural production. The result of the move toward conservation was, as the *True Democrat* editor put it, less land for smaller farms and less cotton cultivation.

Conservation-minded sportsmen also turned their attention to hunting laws to protect access to hunting lands and wildlife. Before examining the role of Hopkins in supporting policy change, an exploration of Red Hills hunting culture—particularly quail hunting—is of use. Coming to Thomas County and the Red Hills region from November to April (when the hunting season ended), sportsmen took advantage of the area's famed shooting. Turkeys, doves, waterfowl, and deer (sometimes even the elusive wildcat) were all prime targets, but the preferred game was the bobwhite quail. The bobwhite quail is a ground-dwelling bird that gathers in coveys of five to thirty birds in the fall and winter months. The Red Hills region is an ideal environment for the bird, which thrives in the brushy edges of cultivated fields, abandoned fields, and long-leaf grassland forests.[47] The pine forests that surrounded Thomasville and the tenant system of labor that scattered farms across the countryside created ideal habitats. One writer to *Forest and Stream* noted the abundance of the bobwhite in Thomas County and the zeal with which "everybody hunts them, both natives and Yankees."[48] He observed that "sometimes one will see a dozen wagons full of men and dogs starting out every morning," to shoot quail.[49] The formal hunting party, with wagons, dogs, and drivers, was the province of wealthy northerners, and quail plantations gave them ample room with which to pursue the practice.

Quail hunting was nothing new to northern or southern hunters, though only during the Gilded Age did it become a formal spectacle. As Nicholas Proctor has argued, small game like quail was often overlooked in the antebellum South among elite hunters in favor of "trophy" animals such as deer and bear that served as symbols of mastery and manhood.[50] Early sporting periodicals, however, attest to the popularity of the bird, at least among northern sportsmen; writers gave much attention, for example, to the natural history, habitat, and behavior of quail. One northern writer considered it a "bird of value" because of its intelligence and the skill, firearms, and dogs required to bag the bird.[51] Hunters in mid-nineteenth-century Illinois, according to one historian, "agreed that quail was the most desirable game and the most difficult to kill on the wing."[52]

Gilded Age quail hunts bore little resemblance to their antebellum predecessors. By the century's end, quail shooting had acquired a pageantry that involved thoroughbred dogs and horses, wagons, and a cast of servants. Sociologist Stuart A. Marks, in writing about hunting in North Carolina, has attributed this transformation specifically to Thomasville, arguing that "the purchase of Southern plantations by Northerners and their use as retreats . . . perpetuated the image of quail hunting as a recreation for the leisured and privileged classes."[53]

Quail hunting, in its elite form, centered on the Georgia hunting wagon and formal hunting parties. Developed in Thomasville, the Georgia wagons (still used for hunting today) have high wheels to enable smooth running through tall grass and brush in the open field and, as Hopkins described to hunter D. L. Hebard, "have to be of extra long bodies" to accommodate the "boxes on sides for guns" and "dog crates in rear" (that hold from four to ten dogs and a water tank).[54] These wagons, pulled by mules, made wainwrighting a lucrative enterprise in Thomasville; Hopkins informed Hebard that in 1930, when wagons were still in demand, they cost around $350. The wagon's accouterments allowed for the socializing that came with formal hunts; Hopkins noted that "lunches, ice, liquid refreshments, etc." were kept in a dash compartment for the midday meal.[55] At that point, the driver(s) would unpack lunch for the hunters, who would linger at the picnic for an hour or so before returning to the hunt or heading home. Hunters, who might ride atop the wagon and or follow on horseback, were accompanied by the wagon drivers and dog handlers, who were also sometimes on horseback.

This type of hunting party differed markedly from most local hunters who pursued game on foot with a single dog and years-worn gun. The observations of Grady C. Cromartie, whom Clifton Paisley interviewed for his work on the Red Hills in 1970, illustrate the kind of spectacle northern hunters created in rural Georgia and Florida. In 1908, Cromartie was clerk at a store that served the farming community surrounding Lake Iamonia in Leon County, Florida, the southern end of the Red Hills region. When asked if he remembered the northern hunter Edward Beadel, who owned a quail plantation on the north side of the lake, Cromartie remarked on Beadel's hunting wagons that were "almost always painted yellow," including the wagon wheels. Cromartie also remembered that Beadel's wagon driver, a black man, "had to be dressed like they wanted him to be dressed," with "leather lines" and formal livery—the driver "had to go neat, don't you know." With day-glo wagons and uniformed drivers, Cromartie noted wryly that northern hunters "were kind of particular . . . about how everything looked."[56]

Conspicuous consumption had a presence in other hunting rituals. Northern hunters also used their lands to host foxhunts, formal affairs that William Rogers notes, "were replicas of similar events in the North and in England."[57] Plantation-based foxhunts featured large packs of dogs, riders in formal habits, and a crowd of spectators. Hunts took place each season, often on the plantations of J. Wyman Jones and Charles Chapin.[58] These sportsmen had acquired so much acreage that riders could follow the baying hounds without leaving their own property. The plantation lands, perfect for quail shooting and large fox chases, served as a backdrop upon which northerners acted as would-be gentry.

The consolidation of lands in the Red Hills region gave elite northern hunters access to the best hunting grounds while simultaneously changing traditional southern hunting practices in important ways. Sportsmen built their estates just as game laws became more stringent and, as some argued, reflective of the interests of elite hunters. Indeed, concerns about game depletion and overhunting spurred northern landowners to begin conservation initiatives on their own plantations. H. W. Hopkins, who occupied the statehouse intermittently from the 1890s to second decade of the twentieth century, used his influence to change state game laws to reflect the growing concern for conservation. Hopkins and others followed national trends in calling attention to the problem of declining game populations. Not the cause of the "true sportsman," who followed bag limits on principle, the culprits of overshooting were "pot-hunters," or those who hunted game for the market, and "game hogs," those men whose kill knew no boundary. It was against these two groups that *Forest and Stream* railed—the rhetoric of Grinnell and others in this regard became common among conservation-minded hunters and anglers and influenced the creation of the Georgia Department of Game and Fish in 1911.

Hopkins became such an important advocate for conservation that in 1915 the commissioner of the Department of Game and Fish called him "one of the best friends of game protection in Georgia" and invited him to use the Department's offices as headquarters for his senate term.[59] In fact, Hopkins' voting record in the Georgia Assembly suggests he was part of a wave of reform that swept through Georgia and the South in the first and second decades of the twentieth century.[60] Some of the reform measures followed Progressive agendas enacted elsewhere in the country.[61] As a state congressman in 1911 he voted to limit working hours for factory laborers, for example, and to allow women to enter the Georgia bar (the latter bill did not pass).[62] He focused on municipal reform in Thomas County and led a statewide effort to tighten prohibition laws in a 1917 special session of the House.[63] And as an avid sportsman and representative of a district that benefitted mightily from northern game hunters, he focused on the creation of more stringent game law.

Hopkins' first run as a statesman was in 1894 and 1895 when he was elected to the House, then again in 1896 when he became a senator. During that time he did not create legislation regarding hunting,[64] though he led the effort to amend a state game and fish law that had come to the Senate from the House (statewide laws began to appear sporadically in the late 1870s).[65] In his later term as a senator between 1902 and 1904, he created restrictions in Thomas County that made it illegal to hunt or fish on another's property without written consent.[66] In doing so, Hopkins was following the lead of other state lawmakers who had passed similar regulations in their own districts. The timing of the measure coincided with Thomas County's rise as a hunting destination for northerners.

Hopkins' efforts figured in a burst of regional game legislation in the late nineteenth and early twentieth centuries, though hunting and fishing law in the South well predates the onrush. In Georgia, for example, game regulation of a kind went as far back as the colonial era, when in 1790 the statutes outlawed hunting deer at night by firelight. In the mid-nineteenth century, lawmakers from individual counties established a hunting season for deer and outlawed the poisoning of fish by dumping walnut hulls or buckeyes into streams.[67] Representatives also passed laws that protected terrapins and oyster beds—particularly beds of the individual property holder—and outlawed camp hunting in coastal Georgia counties.[68] Richmond County set a season for quail, turkey, snipe, ducks, and other wild birds between October 1 and April 10.[69] These piecemeal efforts reflected concern for diminishing game populations and efforts to protect the property rights of landowners. Because these laws remained largely unenforced, however, customary hunting practices, which gave hunters access to unfenced lands, still held sway.

In the first decade of the twentieth century, as suggested by Hopkins' measure, the passage of hunting law remained piecemeal and local. Although state laws regulated seasons for certain types of game and outlawed certain hunting practices, no official body existed to enforce them. Supporters of game law, mostly sportsmen from across the state, argued that haphazard codes and puny enforcement made regulations ineffective. Sportsmen advocated for creation of an agency dedicated to protection of game and fish. In a 1908 letter to the *Columbus Enquirer-Sun*, writer R. Andrews called the warden system in the state "worse than a farce" and compared Georgia's "unique and unenviable" position of being without a game commission with other southern states (except Florida and Mississippi) that did.[70] The next year, the *Atlanta Constitution* issued another call for a game and fish commission. "Constant complaints of the ineffectiveness of Georgia's game law and the known scarcity of wild animals, birds, and fish," the editor argued, "sufficiently evidence the need in this state for a statute which will adequately protect this one of our rapidly diminishing natural resources." He lamented that, because of overshooting, there remained "few spots in the state to which the true sportsmen can go and enjoy a reasonably satisfactory day's outing." "True sportsmen," he noted, were rallying by way of petition and lobby to influence the state legislature to pass a comprehensive game bill, which had been attempted in the 1908 legislative session but had failed.[71]

The swell of support for a game law broke ground when, in 1911, the assembly codified hunting law in the state. In that year, H. W. Hopkins rejoined the House of Representatives and took an active role in efforts to create a department for game and fish regulation. Hopkins and two other congressmen introduced a bill to create a Department of State Conservation. Though it is not clear what the measure entailed or how it differed in substance from other acts to create a unified department, the effort was tabled. Instead, a bill was passed to create the Department of Game and Fish.[72] The Department had powers to appoint a state Game and Fish Commissioner, to select wardens and deputy wardens, to create a licensing structure for in-season hunting, as well as to criminalize violations of game and fish law.

By 1911, those state laws went beyond the establishment of seasons, the outlawing of trapping, and bag limits (which, for quail, were twenty-five birds per diem), all restrictions that were in place by the 1890s. New laws, in effect August 21, 1911, and given "teeth" by the creation of the Department of Game and Fish, made licenses necessary, for example. To hunt in one's own county cost one dollar, in the entire state, three dollars, and, for a non-resident, fifteen dollars. The law did allow that, "a person may hunt and fish in the open season in his own militia district or on his own land without a license," and made it legal for tenants to hunt without a license on "leased and rented" land with the owner's permission. New laws also made it illegal to transport game to another state or county unless accompanied by the game hunter, an amendment squarely aimed at criminalizing market hunting and subsequent sale of game.[73] These laws, informed as they were by men such as Hopkins, followed the statutes of sporting culture elsewhere in the United States.[74] Ostensibly for the conservation of game, new laws privileged hunting for leisure and made operation more difficult for those who hunted for sustenance or additional income. Hopkins, as an avid sportsman and ally to northern hunters in Thomas County and the Red Hills region, and others infused state game laws with the priorities of elite interests. The laws protected property owners and encouraged landholders to post their land, unless the huntsman was following a pack of hounds in chase of a fox or deer and then could trespass freely. The exception to the rule of posted lands privileged those men who had the means to keep hounds for sport. License fees also made hunting accessible only to those who could afford the cost. All of these laws—outlawing the sale of game, posting property, and license fees—went against traditional hunting practices and would ultimately upset small farmers who claimed mastery not only of their lands but also the game therein.

If elite hunters benefited from the new laws, the inverse was true for some whites and most blacks. New legislation not only circumscribed the movement of "pot hunters" and "game hogs" but also black men. Just as the vagrancy and so-called anti-enticement laws of the late nineteenth century had sought to reestablish control over the labor of black men, the new laws restricted activities outside of working for white men and sought to keep firearms out of black hands. After emancipation, no laws barred freedmen from owning guns outright but conservation laws could, if enforced, perform that task.

Farmers and sportsmen saw the potential of game laws to restrict black mobility. Although the code allowed that anyone could hunt in their own militia district without license and that tenants could hunt and fish on land they farmed, permission of the owner permitting, some smallholders, such as Thomas County local R. R. Redfearn, supported the curtailing of even these rights. He wrote to Judge Hopkins concerned about the need for shooting and fishing restrictions because of "triflen [*sic*] negroes" who fished anywhere they pleased.[75] Though Redfearn may not have known it, the State of Georgia was on his side.

Only two months after the assembly created the Department of Game and Fish, the *Atlanta Constitution* reported that the new laws were being used as a "club" against black laborers. Jesse Mercer, the first game commissioner, told the paper that white farmers in Dekalb County reported falsely that "negroes had quit work in the fields and had gone to hunting birds and other game without license and that a ruthless slaughter of all kinds of was being carried on. . . ." Game wardens responded and found no such slaughter had taken place. Farmers used their presence, however, to "spread the report that the game wardens were after the negroes in order to get them back on the farms, where their labor is needed."[76]

Sportsmen also argued for the need to control "indolence" among blacks by curtailing hunting privileges. One common critique of the new game code concerned its allowance of hunting without a license in one's own militia district. This provision frustrated wardens. Militia districts tended to be poorly defined and violations could be easily challenged in courts. Game commissioners consistently advised the Georgia Assembly to change this statute in the code and require a license to hunt anywhere outside of personally owned property. Some commissioners attempted to be more persuasive than others in their reports. Charles Arnow, the Commissioner of the Department of Game and Fish in 1917, wrote in his annual report that "many irresponsible persons" took advantage of the militia district exception and disregarded "other rights of neighbors" who might have posted their land. He singled out "large numbers of negroes" for these types of transgressions, "who are glad of an excuse to prowl through the woods, with dog and gun, when they had far better devote their time and energies to pursuits more in line with their temperaments and necessities." Arnow considered the mobility of black hunters, who bad legal right to hunt without a license, dangerous. To keep black farmers "in line" and at work, he recommended amending the game law to make hunting without a license legal only for property holders.[77] Sam Slate, the Commissioner in 1918, echoed Arnow's recommendation in his report the next year.[78] Game laws enacted in the Progressive Era in Georgia served as a form of social control.

The move to limit black access to hunting also suggests that the Department of Game and Fish, sportsmen, and white farmers equated hunting with a certain kind of manhood; the control over land and game was at the heart of their constructions of elite white manhood or of the independent small farmer. If black men were allowed to possess guns (privileges that had been curtailed while enslaved) and have access to hunting grounds, the mastery and manhood of white men stood to be challenged. The many complaints about black hunters after the Civil War stood as a common "trope of lost control" over a racial hierarchy.[79]

Northern hunters took advantage of the new laws to support and further changes that they had put into practice already in Thomas County and the Red Hills region. Charlie Young, a black man from the area who spent his life working as a dog handler and gardener for northern hunters, was an apt observer of changing hunting practices brought by northern sportsmen. He noted, for example, the premium on quail after the northern hunting colony was formed. When he was a boy in the late 1880s, "Quails was every Whair [*sic*] no one ceard [*sic*] about them" and locals could buy the birds at the market for fifteen cents. Black hunters, Young remembered, could not afford the commonly used shells to hunt quail and certainly couldn't afford the smokeless shells that northern sportsmen introduced to the area. But, they could "set a trap eny Whair [*sic*]" and sometimes bag a whole covey of quail (often fifteen birds). Trapping was outlawed in Thomas County by the state legislature in 1876; the observations of Young suggest that these laws were ignored.[80] But, as land was increasingly privatized and laws enforced by newly empowered game wardens after 1911, the trapping and hunting of local blacks and whites was curtailed in a major way.

Northern hunters also wasted no time posting their lands. Young recounted that signs began to dot the rural landscape, warning trespassers of the consequences if caught poaching. Young recalled, "the first sign put up in this country was by a northern man Dr. Metcalfe," the early proponent of Thomasville who owned a 1,200-acre estate near the town.[81] Northerners also fenced their property. In 1904, just as he purchased land in the area, J. H. Wade wrote to Hopkins that, "I am going to fence my entire property at once. The two ten acres [*sic*] pieces on the Boulevard I would like in order to avoid fencing around them. What can you get them for? I will fence around the Girtman place west of the road and run the present road around it also."[82] A. H. Mason, owner of Susina Plantation, took similar action. In 1910 he purchased 1,000 posts from a local man in order to fence in his lands.[83]

Sportsmen often hired their own wardens to ward off illegal hunters. Efforts to employ a warden sometimes took collective form or benefited from information by multiple landowners in the hiring process. Sometimes a group of plantation owners shared a warden, as was the case with the Lake Iamonia Hunting Club. The club, like its predecessor the Cracker Gun Club, was a collective of wealthy northern hunters and a few of the southern elite, including Hopkins (as president), who shared shooting access to exclusive lands. The club's bylaws provide a clause to employ a game warden that would protect lands from trespassers.[84] In later years, the game warden became a position appointed by the Department of Game and Fish, but Hopkins and the northerners continued to hold sway over appointments in the area. In 1933, C. D. Jordan, from nearby Monticello, Georgia, wrote to Hopkins seeking the position of game warden in Thomas County, and asked that he "write Mr. Lou [*sic*] Thompson and the other millionaire owners of estates down there and ask them to endorse me [to?] Governor Talmadge."[85] Hopkins, who was well connected in the Department of Game and Fish, likely had no problem securing the appointment if he favored Jordan.

The game laws of 1911 and early conservation advocacy did not go unchallenged. Small farmers, who likely approved the effects on African American hunters, had no time for the challenges that new laws gave to their assumed mastery. Farmers from across the state took notice of the state's nod to sportsmen. Some wrote to *The Jeffersonian*, the magazine of famed Populist Tom Watson, to air their grievances. Even before the Department of Game and Fish was in place, W. L. Dorris cast the interests of sportsmen and farmers as being at odds. While sportsmen, he argued, came during hunting season "to the different railroad towns" to shoot indiscriminately, even on posted lands, farmers and tenants who had spent the year raising the birds for meat and for the eradication of insects had to stand by "indignant." Dorris opposed a state licensing program; he argued that "under a State license the State is their domain, and the farmers must stand by and see their birds shot down and their crops trod down, without recourse." Dorris' descriptions of sportsmen, with their "imported setters and pointers and brand new guns that glistened in the November sunshine like so many mirrors," emphatically stressed the moneyed aspect of sport hunting. He warned that Georgia would be without a truly protective game law as long as the "sporting fraternity" had influence over game policy and the "Legislature enact[ed] laws to meet their hearty approval."[86] Dorris foreshadowed the later criticism of new game laws after 1911.

Another landowner, for example, called for the repeal of all extant game laws in 1916, and took particular issue with the illegality of trapping and marketing game. Citing a property owner's right to kill and sell game on his own land, the writer lamented that only "evils" resulted from the law: that the farmer was "deprived of making money legitimately; those not sportsmen are deprived of the privilege of occasionally eating a little game." Ridiculing concerns about conservation as merely lip service, the writer wondered why "conservation" meant sacrifice by the farmers (when it was they who could raise, trap, and transport birds to other lands, for example) while no one stopped the "'sportsman' killing twenty-five birds" per day, "during the open season for 'sport.'" The writer argued that, "the law operates against the land owner and farmer, and prevents real conservation, and likewise against everybody other than the 'sportsman.'"[87] Three years earlier, when the laws were still new, Francis H. Harris of Brunswick, Georgia, a coastal destination popular with hunters and anglers, lambasted the state laws as protective only of sportsmen. Like the writer in 1916, Harris found the outlawing of killing and selling game found on one's own property to be an egregious violation; he reasoned that if farmers could raise and sell their own stock, they could raise and sell wild game. After Harris' lengthy critique and call for farmers to cry out in protest, Watson agreed that "no man could be deprived of the legal right to protect his crops, at all times," from wild game and birds by trapping and shooting outside of hunting season. "Legislative enactments to the contrary," Watson concluded, "are pluperfect hog-wash."[88]

Local hunters contested the new laws by simply ignoring them. State commissioner reports consistently decried blatant transgressions of game law across the state. Commissioners pleaded with the General Assembly to create more stringent laws to protect wildlife and to establish more equitable compensation for game wardens. In the Red Hills region, new laws and private efforts by northerners did not keep out poachers, which suggests that hunting practices remained largely unchanged. Charles Chapin asked H. W. Hopkins, for example, if he had "heard of any shooting or poaching out around my T.C. Mitchell lands and if so is there anything you could do or I, to stop it." He ventured to Hopkins that, "maybe something could be done to avoid finding birds shot up as I did last year."[89] To mitigate poaching, northern hunters employed game wardens to patrol their properties and remain on the lookout for trespassers. C. A. Griscom, who owned land in Leon County, Florida, wrote to A. H. Mason that he was anxious to find a warden for his lands:

> As far as I know we have no Game-warden yet and I consider the situation precarious. Mr. R.G. Johnson, who is my Agent and lives on Horseshoe Plantation, is trying now to find a man for reasonable wages who has the ability and the nerve. It is no easy position to fill. I will seek your advice if we succeed in securing a suitable man.[90]

Griscom's letter suggests that northern hunters took poaching and trespassing seriously, and his anxiousness indicates that some locals continued hunting as they did before, regardless of posts or fences.

Northerners maintained their close connection with men like Hopkins, who continued to promote their interests locally and statewide, well into the 1930s. Hopkins acted as agent for many northerners up to that point, including L. S. Thompson, John F. Archbold, and George F. Baker of New York.[91] The main purpose of the winter colony—hunting—remained in place during those years and Hopkins and the northerners continued to identify as sportsmen. The leisure economy had, beginning in the late nineteenth century, only deepened the social divides that characterized the South: the continuation of the land tenure system, overwhelmingly populated by black farmers; the struggle by small farmers against the privatization of hunting grounds and consolidation of the area's most fertile lands; and the representation of elite interests in state law, in this case with the creation of the Department of Game and Fish and the game laws codified therein.

Historian Albert Way has chronicled the growth of conservation initiatives in the Red Hills region in the 1920s and 1930s. With the help of naturalist Herbert Stoddard, northerners transformed their hunting plantations into laboratories to study and propagate quail and to develop methods of land management congruent with emerging ideas about conservation. Way argues that Stoddard, in his work on the quail preserves of the Red Hills region, is an important and often overlooked figure in the American conservation movement. Stoddard's ability to bring together the emerging professional and scientific priorities of land resource management and local environmental knowledge and practice—particularly controlled burning—developed a model of "biocentric" management that persists today (several of the game plantations, including Tall Timbers, once home to Edward Beadel's yellow wagons, remain intact as preserves dedicated to research and conservation). But Way concedes that the southern conservation movement headed by Stoddard and the northern estate owners was essentially conservative in nature. Quail plantations always remained in private hands, as opposed to state and federal land trusts, and the movement was founded as "less an oppositional reaction to the growth of industrial and corporate America than a concomitant to it." Ultimately, northern owners "did as they pleased under the property rights structure of the post-open range New South."[92]

Northerners certainly did as they pleased, but relied upon their southern allies to support and further their initiatives. A common identity as sportsmen, one that was cross-regional but class-specific, bound these men together. In the Red Hills region, northerners found an ally in H. W. Hopkins, who did much to construct the northern hunting colony and to see that sportsmen's interests influenced changing game laws. Though sportsmen were successful in realizing their colony and could, by 1911, rely upon a set of laws to enforce their particular hunting culture, they met challenges by local farmers who cast their opposition against a monied elite. Competing versions of manhood lay at the heart of these debates, and though sportsmen were successful in changing policy, small farmers and other non-elite locals were successful in ignoring it, at least until the 1930s. The legacy of northern sportsmen in the Red Hills region is an important one—land preserves founded by them are still in place today—but the history of their seasonal settlement was rife with local challenge. Those challenges must be acknowledged to understand the impact of northern hunting colonies across the region.

NOTES

1. Donald H. *Doyle, New Men, New Cities, New South: Atlanta, Nashville, Charleston, Mobile, 1860–1910* (Chapel Hill: University of North Carolina Press, 1990). Doyle identifies "new men" as urban elites who were invested in the growth of southern cities after the Civil War. These men were lobbyists for railroad expansion, which they linked to economic development, and saw opportunities for a growing urban network linked to industrial development.

2. For more detail about small farmers' conception of manhood, see Stephen Kantrowitz, *Ben Tillman and the Reconstruction of White Supremacy* (Chapel Hill: University of North Carolina Press, 2000), esp. chap. 4.

3. Albert G. Way, *Conserving Southern Longleaf: Herbert Stoddard and the Rise of Ecological Land Management* (Athens: University of Georgia Press, 2011), p. 12.

4. Clifford Paisley, *From Cotton to Quail: An Agricultural Chronicle of Leon County, Florida, 1860–1967* (Gainesville: University of Florida Press, 1968), p. 77.

5. A vast literature on the anthropology of tourism seeks to define tourism and the tourist. This literature is useful for historical studies of tourism. I use a broad definition that does not distinguish between different subsets of tourism for business, health, recreation, and so on. Consequently, I am less concerned with discerning the motivations of early tourists to Thomasville than assessing their influence. For classic studies of the anthropology of tourism, see Dean MacCannell, *The Tourist: A New Theory of the Leisure Class* (Berkeley: University of California Press, 1976); Victor Turner and Edith Turner, *Image and Pilgrimage in Christian Culture* (New York: Columbia University Press, 1978); John Urry, *The Tourist Gaze: Leisure and Travel in Contemporary Society* (London: Sage Publications, 1990).

6. Paisley, *From Cotton to Quail*, p. 77.

7. William R. Brueckheimer, "The Quail Plantations of the Thomasville-Tallahassee-Albany Regions," *Journal of Southwest Georgia History* 3 (fall 1965): p. 44.

8. Nicolas Proctor, *Bathed in Blood: Hunting and Mastery in the Old South* (Charlottesville: University of Virginia Press, 2002), p. 61.

9. Ibid., p. 145. See chap. 7 for a more thorough look at the hunting experiences of enslaved men.

10. "With Gun and Dog in Georgia," *Forest and Stream* 60, no. 2 (Jan. 10, 1903): p. 29.

11. E. L. Youmans, "Thomasville as a Winter Resort," *Popular Science Monthly* 28 (Dec. 1885): p. 190.

12. "Beagles and Bulls and Other Attractions of Thomasville," *Atlanta Weekly-Constitution* (Atlanta, Ga.), May 2, 1882, p. 7.

13. *Forest and Stream* 18, no. 18 (June 18, 1882): p. 376.

14. "Beagles and Bulls and Other Attractions of Thomasville."

15. Proctor, *Bathed in Blood*, p. 31.

16. Daniel Justin Herman, *Hunting and the American Imagination* (Washington, D.C.: Smithsonian Institution Press, 2001), p. 139.

17. Ibid., p. 222.

18. D. L. Shepard to H. W. Hopkins, Apr. 26, 1901, fol. 1978.010.188, box 3010A, Hopkins Collection (hereafter HC), Thomas County Historical Society (hereafter TCHS).

19. C. S. Hebard to H. W. Hopkins, July 14, 1903, fol. 1978.010.269, box 3019A, HC, TCHS.

20. William Warren Rogers, *Thomas County 1865–1900* (Tallahassee: Florida State University Press), p. 116.

21. Brueckheimer, "The Quail Plantations of the Thomasville-Tallahassee-Albany Regions," p. 53; William Warren Rogers, *Antebellum Thomas County 1825–1861* (Tallahassee: Florida State University Press, 1963), pp. 67, 115.

22. Brueckheimer, "The Quail Plantations of the Thomasville-Tallahassee-Albany Regions," p. 54; Rogers, *Thomas County*, p. 116. Susina remained in the Mason family until 1980; today it is still a private residence.

23. Paisley, *From Cotton to Quail*, p. 78; "On the River," *Thomasville Times-Enterprise* (Thomasville, Ga.), Mar. 4, 1893.

24. Brueckheimer, "The Quail Plantations of the Thomasville-Tallahassee-Albany Regions," p. 54.

25. "What a Visitor Says," *Thomasville Times-Enterprise*, Feb. 2, 1893; Paisley, *From Cotton to Quail*, p. 78.

26. Brueckheimer, "The Quail Plantations of the Thomasville-Tallahassee-Albany Regions," pp. 55–56.

27. Ibid. For a full history of Pebble Hill Plantation, see William Warren Rogers, *Pebble Hill: The Story of a Plantation* (Tallahassee: Sentry Press, 1979).

28. William R. Mitchell, Jr., *Landmarks: The Architecture of Thomasville and Thomas County, Georgia, 1820–1980* (Thomasville, Ga.: Thomasville Landmarks, 1980), p. 32.

29. Rogers, *Pebble Hill*, pp. 130–131.

30. J. H. Wade to H. W. Hopkins, Aug. 2, 1904, fol. 1978.010.269, box 3019A, HC, TCHS.

31. J. H. Tayler to H. W. Hopkins, May 6, 1907, fol. 1978.010.269, box 3019A, HC, TCHS. Tayler acted as McCartney's agent.

32. Abstract of Title, "South half of lot number 94 in the 13th District of Thomas County, State of Georgia," H. J. and A. T. McIntyre to J. H. Wade, 1910, fol. 1978.010.269, box 3019A, HC, TCHS; Abstract of Title, "Lot 93 in the 13th District of Thomas County, Georgia," H. J. and A. T. McIntyre to J. H. Wade, 1910, fol. 1978.010.269, box 3019A, HC, TCHS.

33. J. H. Wade to H. W. Hopkins, Oct. 26, 1916, fol. 1978.010.269, box 3019A, HC, TCHS.

34. Journal entry for Mar. 27, 1919, fol. 1:11, Lula Mae Hamilton Harding Collection, Hargrett Library, University of Georgia, Athens, Ga.

35. Paisley, *From Cotton to Quail*, p. 94.

36. H. W. Hopkins to Edward Crozer, Oct. 8, 1915, fol. 1978.010.330–9, box 3020A, HC, TCHS.

37. William Warren Rogers, *Transition to the Twentieth Century: Thomas County, Georgia, 1900–1920* (Tallahassee: Sentry Press, 2002), p. 71.

38. *Cairo Messenger* (Cairo, Ga.), June 24, 1904.

39. Rogers, *Transition to the Twentieth Century*, pp. 69, 72–73.

40. "An Open Letter to Mr. Hopkins," *Thomasville Times-Enterprise*, Feb. 14, 1905, fol. 1978.010.538, box 3012A, HC, TCHS.

41. *Cairo Messenger*, June 3, 1904.

42. *Weekly True Democrat* (Tallahassee, Fla.), July 3, 1914, quoted in Paisley, *From Cotton to Quail*, p. 84.

43. *Weekly True Democrat*, Sept. 24, 1909, quoted in William Warren Rogers, *Foshalee: Quail Country Plantation* (Tallahassee: Sentry Press, 1989), p. 80.

44. *Weekly True Democrat*, Jan. 30, 1920, quoted in Paisley, *From Cotton to Quail*, p. 84.

45. Charlie Young, "Reminiscences of Charlie Young for Bill Rogers," vertical files, accession number 2001.99.104, TCHS, p. 28; Rogers, *Foshalee*, pp. 112–117.

46. Herbert L. Stoddard, *The Bobwhite Quail: Its Habits, Preservation, and Increase* (New York: Charles Scribner's Sons, 1931), pp. 350–351.

47. Way, *Conserving Southern Longleaf*, p. 39.

48. H. C. S., "A Georgia Quail Country," *Forest and Stream*, Feb. 24, 1893, p. 161.

49. Ibid.

50. Proctor, *Bathed in Blood*, p. 58.

51. Samuel B. Smith, "Interesting Particular in the Natural History of the Quail," *American Turf Register and Sporting Magazine*, Dec. 1829, p. 205. See also, "The Quail or Partridge," *American Turf Register and Sporting Magazine*, Jan. 1830, p. 247.

52. John T. Flanagan, "Hunting in Early Illinois," *Journal of the Illinois State Historical Society* 72, no. 1 (Feb. 1979): p. 6.

53. Stuart A. Marks, *Hunting in Black and White: Nature, History, and Ritual in a Carolina Community* (Princeton: Princeton University Press, 1990), p. 71.

54. H. W. Hopkins to D. L. Hebard, Oct. 10, 1930, fol. 1978.010.326–7, HC, TCHS; Paisley, *From Cotton to Quail*, p. 85.

55. H. W. Hopkins to D. L. Hebard, Oct. 10, 1930, fol. 1978.010.326–7, HC, TCHS.

56. Grady C. Cromartie, interview by Clifton Paisley, Jan. 23, 1970, Clifton Paisley Collection, 1915–1968, Special Collections, Archives, and Manuscripts, Florida State University, Tallahassee, Fla.

57. Rogers, *Transition to the Twentieth Century*, p. 215.

58. Ibid.

59. Hopkins served in the state legislator in 1902–1904, 1911–12, 1913–1914, 1915–1916, 1917–1918. "Offices Held by H. W. Hopkins," fol. 1978.010.600, box 3020A, HC, TCHS; Charles H. Arnow to H.W. Hopkins, Oct. 1, 1915, fol. 1978.010.516B, box 3025A, HC, TCHS.

60. For an overview of Progressive law in Georgia, see Numan V. Bartley, *The Creation of Modern Georgia*, 2nd ed. (Athens: University of Georgia Press, 1990).

61. For more on southern Progressivism, see William A. Link, *The Paradox of Southern Progressivism, 1880–1930* (Chapel Hill: University of North Carolina Press, 1997).

62. *Journal of the House of Representatives of the State of Georgia, at the Regular Session of the General Assembly at Atlanta, Wednesday, June 29, 1911* (Atlanta: Charles P. Byrd, State Printer, 1911), pp. 706, 803.

63. See *Journal of the House of Representatives of the State of Georgia, Extraordinary Session at Atlanta Tuesday, March 20, 1917* (Atlanta: Byrd Printing, 1917).

64. He amended dates on the hunting season and took out snipe as a protected game bird. See *Journal of the Senate of the State of Georgia, at the Regular Session of the General Assembly, At Atlanta, Wednesday, October 28, 1896* (Atlanta: George W. Harrison, 1896), pp. 355–356.

65. Lawmakers passed bills in 1878, for example, that allowed landowners to post their lands to stop trespassing hunters and that outlawed trapping fish in creeks and rivers. See *Acts and Resolutions of the General Assembly of the State of Georgia, 1878–1879* (Atlanta: James P. Harrison & Co., 1880), p. 52.

66. *Journal, the Senate of the State of Georgia, at the Regular Session of the General Assembly, At Atlanta, Wednesday, June 24, 1903* (Atlanta: Franklin Printing Publishing Co., 1903), p. 233.

67. John Rutherford, comp., *Acts of the General Assembly of the State of Georgia Passed in Milledgeville at a Biennial Session, in November, December, January, and February, 1853–54* (Savannah: Samuel T. Chapman, 1854), pp. 336–340.

68. John W. Duncan, comp., *Acts of the General Assembly of the State of Georgia, Passed in Milledgeville at a Biennial Session, in November, December, January February, and March, 1855–56* (Milledgeville: Boughton, Nisbet and Barnes, 1856), p. 12.

69. *Acts of the General Assembly of the State of Georgia, Passed in Milledgeville, at an Annual Session in November and December, 1859* (Milledgeville: Boughton, Nisbet, and Barnes, 1860), p. 9.

70. R. Andrews, "State Without Game Law—Georgia's Unique and Unenviable Distinction," *Columbus Enquirer-Sun* (Columbus, Ga.), reprinted in *Atlanta Constitution* (Atlanta, Ga.), June 29, 1908, p. 4.

71. "Georgia Needs an Effective Game Law," *Atlanta Constitution*, Feb. 22, 1909, p. 4.

72. *Journal of the House of Representatives of the State of Georgia at the Regular Session of the General Assembly* (Atlanta: Charles P. Byrd, 1911), July 10, 1911, p. 284; July 14, 1911, p. 427; and *Acts and Resolutions of the General Assembly of the State of Georgia, 1911* (Atlanta: Charles P. Byrd, 1911), Part 1, Title 5, pp. 137–146.

73. Charles B. Reynolds, *The Game Laws in Brief: A Digest of the Statutes of the United States and Canada Governing the Taking of Game and Fish* (New York: Forest and Stream Publishing Co., 1911), p. 89, accessed Aug. 20, 2011, http://www.archive.org/stream/gamelawsinbrief00canagoog#page/n5/mode/2up.

74. Scott Giltner, *Hunting and Fishing in the New South: Black Labor and White Leisure After the Civil War* (Baltimore: Johns Hopkins University Press, 2008), pp. 150–158.

75. R. R. Redfearn to H. W. Hopkins, June 20, 1903, fol. 1978.010.515–5, HC, TCHS.

76. "State Game Law is Used as Club: Farmers Take Advantage of Law to Frighten Negroes," *Atlanta Constitution*, Nov. 30, 1911, p. 6.

77. Charles Sterling Arnow, *Sixth Annual Report of the Department of Game and Fish of Georgia, July 1st 1916 to June 30th, 1917, State Game and Fish Commissioner* (Atlanta: Johnson-Dallis Co., 1917), p. 5.

78. Sam J. Slate, *Seventh Annual Report of the Department of Game and Fish of Georgia, July 1st, 1917 to June 30th, 1918, State Game and Fish Commissioner* (Atlanta: Johnson and Dallis Co., 1918), p. 5.

79. Giltner, *Hunting and Fishing in the New South*, p. 51.

80. Rogers, *Thomas County*, p. 259.

81. Young, "Reminiscences," p. 26.

82. J. H. Wade to H. W. Hopkins, Aug. 2, 1904, fol. 1978.010.269, box 3019A, HC, TCHS.

83. "Memorandum of Accounts Paid by H. W. Hopkins for Account, A. H. Mason, From Sept., 25th, 1909 to May, 1910," fol. 1978.010.320–11, box 3020A, HC, TCHS.

84. Lease for the Iamonia Lake Club, n.d., fol. 1978.010.611, box 3010A, HC, TCHS.

85. C. D. Jordan to H. W. Hopkins, Aug. 2, 1933, fol. 1978.010.518, box 3025A, HC, TCHS.

86. W. L. Dorris, "Discussing the Game Laws," *The Jeffersonian*, Apr. 19, 1909, pp. 6–7, Thomas E. Watson Papers Digital Collection, Southern Historical Collection, University of North Carolina, Chapel Hill, accessed Aug. 19, 2011, http://www.lib.unc.edu/dc/watson/index.html/container/digitem_142?id=006.

87. A Georgia Citizen, "Interesting Facts About the Game Laws," *The Jeffersonian*, Feb. 3, 1916, accessed Aug. 19, 2011, http://www.lib.unc.edu/dc/watson/index.html/container/digitem_405?id=003.

88. Frances H. Harris, "Criticizes the Georgia Game – Law," *The Jeffersonian*, Mar. 6, 1913, p. 9, accessed Aug. 19, 2011, http://www.lib.unc.edu/dc/watson/index.html/container/digitem_288?id=009.

89. Charles M. Chapin to H. W. Hopkins, n.d., fol. 1978.010.317, box 3020A, HC, TCHS.

90. C. A. Griscom to A. H. Wade, Nov. 8, 1905, fol. "Hunting Rights, Sale of Plantations," box 3005A, HC, TCHS.

91. H. W. Hopkins to C. M. Chapin, Jr., July 6, 1933, fol. 1978.010.416–4, box 3020A, HC, TCHS.

92. Way, *Conserving Southern Longleaf*, p. 226.

Chapter 7

Life and Labor on the Southern Sporting Plantation

African American Tenants at Tall Timbers Plantation, 1920–1944

Robin Bauer Kilgo

In the decades following the Civil War, sharecropping and tenancy quickly replaced slave labor on plantations across the South. Sharecropping and tenancy mediated the conditions of the immediate postwar era by satisfying landowners' desire to restart production, freedpeople's desire for autonomy, and the difficulty of paying regular wages in a cash-poor society. By the early 1880s, both systems had taken root across large portions of the former Confederacy. In the years that followed, they became widespread. By shaping the lives of millions of southerners, white and black, sharecropping and tenancy became central to the social, economic, and cultural fabric of the post-emancipation South.[1]

Historians long portrayed sharecropping and tenancy as debilitating institutions that trapped landless farmers in debt peonage with little chance of escape. In his history of postwar southern agriculture, Gilbert Fite characterized their rise as a "descent into poverty." Contemporary observers such as Frank Tannenbaum credited both institutions with "pauperizing" the South. Throughout the 1920s, 1930s, and 1940s, policymakers, social scientists, and politicians saw sharecropping and tenancy as grievous problems, fundamental to the social and economic ills that plagued the southern states.[2] These judgments, however, rest entirely on plantations operated as agricultural enterprises. How did conditions differ on plantations devoted to sport and leisure? Did tenants experience better relations with landowners, or did the difference in land use make a negligible difference in their lives?

Tall Timbers Plantation affords an exceptional opportunity to examine these questions. Established at the end of the nineteenth century by Edward Beadel, heir to a New York real estate fortune, Tall Timbers lies in northern

Leon County, Florida, on the edge of Lake Iamonia. Like most sporting plantations in the Red Hills region, its lands had previously grown cotton and other crops. Tenant farming became well established in Leon County after the Civil War. Beadel modified the conditions of tenancy at Tall Timbers but did not alter them radically. Continuity more than change characterized the transition from agriculture to sporting use.

The story of the African American tenancy at Tall Timbers might have remained unexplored had it not been for the restoration and public interpretation of a surviving tenant homestead. An unusually rich array of sources offers insight into the lives of tenants at Tall Timbers. Archaeological investigations, extant buildings, plantation ledgers, and oral histories supply valuable information. Together, these sources yield a portrait of tenancy with an accommodating landowner and modestly better material conditions than commonly found on agricultural plantations. Although manifold inequalities existed, the case of Tall Timbers suggests the potential for studies of sporting plantations to revise the traditional portrait of tenancy in the early twentieth-century South. The priorities of sportsmen and sportswomen and their labor needs led to arrangements that offered tenants somewhat greater autonomy, security, and stability. Their lives remained difficult and bereft of opportunity. Still, compared to other people near the bottom rung of rural southern society in the era between the world wars, tenants at Tall Timbers fared better than many.

TALL TIMBERS: ORIGINS AND DEVELOPMENT

Tall Timbers Research Station is an ecological research center on the northern outskirts of the Tallahassee metropolitan area, immediately south of the Georgia-Florida border. Founded in 1958, Tall Timbers embodies the legacy of Henry Beadel's love of hunting and the Red Hills landscape. Beadel purchased Tall Timbers from his uncle, Edward Beadel, in 1919. He subsequently became deeply involved in land and wildlife conservation. In the early 1920s, Red Hills landowners became alarmed by a rapid decline in the quail population. They responded by sponsoring a scientific investigation of the problem in collaboration with the U.S. Biological Survey. Herbert L. Stoddard, a naturalist and ornithologist, conducted the investigation. He quickly determined that the form of forest conservation advocated by federal authorities, which emphasized natural regeneration and saw fire as destructive, had caused the loss of quail habitat. Stoddard recognized that man-made patterns of land use created a patchwork landscape with ample nesting and feeding areas. Farming, forestry, and controlled burning all proved beneficial. Stoddard published his findings in 1931 as *The Bobwhite Quail: Its Habits,*

Preservation, and Increase. The book established him as a leading authority of wildlife management. The success of Stoddard's investigations led
Red Hills landowners to found the Cooperative Quail Study Association, an
organization dedicated to establishing proper game-management practices
on sporting retreats across the South. Stoddard served as its initial director.[3]

Henry Beadel enthusiastically supported Stoddard's work. A naturalist
in the idiom of Teddy Roosevelt and other gentleman hunters of the Gilded
Age, Beadel visited Leon County every winter from 1894 until purchasing
Tall Timbers.[4] He honed his interest in wild animals, birds especially, at
the Cedars, his family's 36-acre estate on Staten Island. After 1919, Beadel
became a Florida resident. An architect by training, Beadel practiced until
the mid-1940s, when he turned his attention to studying wildlife and hunting.
He became renowned for his photographs and film footage of birds at Tall
Timbers, which today form one of the finest collections of ornithological
images and films in the Southeast.

Beadel made provisions in his will for Tall Timbers to become an ecological research center upon his death. It has operated in this manner since the
early 1960s. Quail is no longer hunted at Tall Timbers but, rather, tagged for
study by trained biologists and conservation specialists. Tall Timbers also
continues to investigate the use of fire as a land-management tool and has
become a clearinghouse of information about game birds, forest ecology, and
wildlife management. In addition, Tall Timbers advocates conservation stewardship of property in the Red Hills region and neighboring areas. Its efforts
have protected more than 160,000 acres through conservation easements.[5]

The land encompassed by Tall Timbers first entered agricultural use by
whites in the 1820s, as settlers poured into the Florida Territory and short-
staple cotton cultivation spread westward into untapped territory. In the
mid-1830s, Griffin Holland, a doctor from Virginia, established Woodlawn
Plantation. By 1840 he had fifty-three slaves working his lands. By 1860 his
labor force numbered 105 men, women, and children ranging in age from a
few months to more than 100. Most worked as field laborers. Holland housed
his slaves in twenty cabins at Woodlawn. He owned more than 2,500 acres,
of which 1,200 were improved. In 1860, Woodlawn produced 225 bales of
cotton and 7,000 bushels of corn.[6]

Woodlawn's output illustrates the growth of plantation agriculture in Leon
County, which ranked as Florida's wealthiest county during the late antebellum era. More than fifty planters owned more than fifty slaves each. Yeoman
farmers also prospered. Slaves made up more than three-quarters of the
county's 12,343 inhabitants by 1860. In that year, planters produced 16,686
bales of ginned cotton, the largest harvest recorded in Florida.[7]

The Civil War and emancipation triggered sweeping upheavals in Leon
County agriculture. "The plantations are mostly waste," observed the

Semi-Weekly Floridian in 1867. Planters increasingly concentrated on cotton but faced declining prices and difficulty obtaining labor. Although planters sought to return to the gang-labor system used before the war, freedpeople refused. Sharecropping and tenancy developed rapidly as a compromise to sharply differing interests. By the early 1870s, these arrangements and wage labor sustained agricultural production in Leon County.[8]

Holland's fortunes plummeted during the Civil War era. He lost nearly half of his personal wealth in the 1860s and returned to Virginia. In 1871, he sold Woodlawn to Alexander Moseley, a Leon County native and Confederate veteran.[9] Mosely, then thirty-one, operated Woodlawn for a decade before becoming sheriff of Leon County. In 1880 he sold the plantation to Eugene H. Smith, a merchant from Thomasville, Georgia, for $4,000. Smith renamed his new property Hickory Hill and took up planting with vigor.[10] Tenant farming flourished under his tenure. Members of the Nix, Wyche, and Stratton families lived on his lands. Some may have been born as slaves on Holland's plantation; others had arrived in 1865–1867, when homesteading opportunities and strong demand for plantation laborers brought nearly 5,000 African Americans to Leon County.[11]

In 1895, Edward Beadel purchased Hickory Hill for $8,000. An avid quail hunter, Beadel did not intend to operate an agricultural enterprise. He renamed the plantation Tall Timbers and immediately began turning it into a hunting estate. His activities fit within a growing trend. The Red Hills region, which occupies portions of Thomas County, Georgia, and Leon County, became a prized destination for quail hunters during the late nineteenth century. Many hunters came from northern cities. Depressed land prices facilitated acquisition of large tracts. By 1900, wealthy northerners had purchased at least 10,000 acres in Leon County. Across the border in southern Georgia, other sportsmen also purchased land.[12]

Beadel left the tenant farming system in place. Small-scale farming, as Stoddard later demonstrated, created favorable conditions for quail and other wildlife. Quail especially thrive in the brushy edges that typically border small farms; Edward Beadel recognized the ecological value of small-scale farming and saw no reason to change existing practices. An enthusiastic hunter, Beadel had previously spent winters near Thomasville and hunted extensively in the Red Hills. He operated his plantation in a manner consistent with other sporting estates.[13]

When Henry Beadel purchased Tall Timbers from his uncle in 1919, he instituted a number of changes. He continued to use Tall Timbers as a sporting plantation but renovated and enlarged the main house and made improvements to several outbuildings. The main house is a two-story frame building set on brick piers. Built circa 1895, it offered comfortable but unremarkable accommodations. Beadel carried out extensive repairs, added a t-shaped wing containing a master bedroom, and modified the main porch. These changes gave the building a more handsome appearance and greater interior space. The house occupies a gentle slope overlooking Lake Iamonia, giving it a panoramic view of the lake and surrounding countryside. Nearby stand several buildings erected by Edward Beadel circa 1895, including a cook's house, a pump house, and a hay barn. Henry Beadel added a dairy barn and corncrib sometime during the 1920s.[14]

Henry Beadel employed a combination of share and cash tenant contracts at Tall Timbers. He recorded all of his contracts as cash transactions but modified them as needed, usually when farmers suffered from poor crop yields or encountered other difficulties. Thus, he accommodated tenants' circumstances to some degree. Beadel provided houses for his tenants, in customary fashion for large landowners. Tenants provided their own tools and livestock.[15]

The landscape of tenancy at Tall Timbers assumed the pattern characteristic of plantations across southern Georgia and north Florida. Croplands and tenant homesteads lay interspersed with woodlands in a mixed pattern that afforded tenants privacy and supported Beadel's recreational activities. Beadel maintained a series of irregularly shaped hunting courses that added further complexity and variation.[16]

Beadel's rents varied depending on the size of the farm and market conditions. In 1920 Henry Beadel recorded nine tenant farm contracts on Tall Timbers, three for two-mule farms (fifty-five to sixty acres of land), five for one-mule farms (thirty to forty acres of land), and one for half of a farm (twenty acres).[17] The two-mule farms each rented for $200.00. Three families, the Fishers, the Joneses, and the Gays, farmed them, and each family had between six and eleven members.[18] Smaller families and couples rented and farmed the five one-mule farms at the rate of $100.00 annually. A single man, Cooper Robinson, held the half-farm contract, for which he paid Beadel $50.00.[19]

Beadel's rates remained unchanged until 1926, when the average price per pound of cotton fell from twenty-one cents to twelve cents and the price of corn dropped from ninety-three cents to eighty-four cents a bushel.[20] Tenants shifted production to cotton, corn, sugarcane, peanuts, and sweet potatoes.[21] The boll weevil bore principal responsibility for the drop in prices.[22] The damage it caused, coupled with erosion and soil exhaustion, placed Southeastern farmers in desperate circumstances. The agricultural depression that affected most of the South during the 1920s hit Florida and Georgia farmers especially hard. Tenant farming in Leon County declined dramatically. The 1900 census listed 1,775 tenant farmers, 1,664 of them African American. By 1920, the number of African American tenant farmers had dropped to 1,045.[23] At Tall Timbers, the number of tenant contracts fell from nine in 1920 to seven in 1929. The drop in rental prices that Beadel undertook in 1926 likely stopped others from leaving. The new price structure set the cost of a contract for a one-mule farm at $75.00 and a two-mule farm contract at $112.50. Beadel maintained these prices throughout the 1930s.[24]

Even with the new rental prices, the overall population of tenants continued to decline. In the 1920s, eighty-five people belonging to fourteen different families lived at Tall Timbers.[25] In the 1930s, the population declined to forty-nine members (seven families).[26] Further declines occurred in the following decade, when the overall population dropped to forty-three.[27] Even with the loss of population, the size of tenant farming families at Tall Timbers remained large. In the 1920s and 1930s, family units had an average of six members. During the 1940s the number dropped to five. By comparison, Charles S. Johnson's study of Macon County, Alabama, found average family sizes of five during the 1930s.[28]

Tenant farm families of Tall Timbers mainly consisted of two-parent households. Single men and some single women also contracted with Beadel. Of the thirty-six contracts Beadel recorded from 1920 to 1949, men held twenty-seven, women held six, and men and women held the remaining three jointly. Of the last group, two were husband and wife, while one was a woman and her son (Table 7.1).[29]

Tenants at Tall Timbers maintained somewhat greater stability than proved common elsewhere. Sporting plantations generally supported consistency among tenants because of less coercive landlord-tenant relations. Sociologist Charles S. Johnson found that Macon County, Alabama, tenants moved frequently. Of 612 families, Johnson determined that over half had moved over the course of five years, and a significant number had moved multiple times.[30] A study undertaken by economist E. A. Goldenweiser and demographer Leon E. Truesdell in the 1920s reported similar findings. As they explained, "the most undesirable feature of tenancy in the United States lies in the fact that tenants do not stay long enough on their farms." In 1920s, of 501,748 tenants in the South Atlantic region (including Florida), roughly 44 percent had remained on farms less than two years.[31] By contrast, tenants at Tall Timbers had an average residency length of almost five years.[32] Other studies indicate that tenant farmers on Leon County sporting plantations moved less often because of reasonable rental costs, the availability of game, and favorable farming conditions.[33] The following table enumerates tenant farmers who lived on Tall Timbers between the 1920s and the 1940s and their length of stay.

It is important to note that not all tenants stayed at Tall Timbers for long. The relatively high average length of residence is in part due to the exceptionally long stays of William Gay and Walter Harvin. When these two tenants are excluded, the average period of residence falls to 3.6 years.

Table 7.1   Tenancy Length

Tenant Name	Year(s) in Beadel Ledger	Farming Seasons
Alonzo Bivens	1925–1932	8
Adam Bryant	1920–1923	4
Jim Bob Fisher	1920–1925	4
Johnnie Fisher	1921	0
Julee Fisher	1923	1
Rebecca Fisher	1921–1924	4
Dora Franklin	1944–1948	5
Emmitt Gay	1941–1942	1
Dan Gay	1931–1934	4
William Gay	1920–1949	30
Angeline Green	1922–1924, 1930–1934	3
Angeline and Henry Green	1925–1929	5
Charley Green and Rebecca Fisher	1920–1921	1
William Green	1920–1921	2
Josh Harvin	1941–1948	8
Tom Harvin	1943–1944	2
Walter Harvin	1922–1943	22
Kate and Tom Harris	1921–1923	2
John Hayes	1925–1934	10
Bill Jones	1920–1929	10
Maggie Jones	1930, 1935	2
Lige Jones	1941–1949	9
Lonza Jones	1942–1948	7
Mose Jones	1941–1944	4
Richard Jones	1941	1
Sam Jones	1920–1923	4
Cooper Robinson	1920	1
Mamie Green Smith	1933–1936, 1943–1944, 1946	4
Robert Scott	1931–1933	0
Hattie Strattin	1948	1
Henry Vickers	1924–1925, 1927–1936	12
Richard Vickers	1941–1942	2
John Williams	1937	1
Tom Wilson	1923–1924	1
Ike Witherspoon	1920	1
Mary Wyche	1920	1

Closer analysis of Beadel's records reveals two distinct groups: those who stayed for several years and those who did not. Of the thirty-six tenants listed, seventeen (47 percent) remained for two seasons or less. Nearly as many—sixteen—stayed between three and twelve seasons.

By the 1940s, nine African American families farmed as tenants at Tall Timbers. Two of these families had roots in the first generation of tenants that worked the land; the others had arrived more recently. The majority of the latter group had lived in the surrounding area for some time.[34] Although tenant

farming employed the largest percentage of African Americans on plantations, others worked as wage laborers for the owner. Sportsmen employed
blacks as support staff in running the main house and in hunting expeditions.

In a significant number of cases, members of the same families rented land
from Beadel and worked for him as wage laborers. Henry Vickers, for example, rented a land from Beadel throughout the 1920s and 1930s before becoming a wage laborer in the 1940s, when he earned $25.00 a month. While
Vickers worked for Beadel, his son, Richard, tried his hand at farming for at
least two seasons in the early 1940s. He rented land from Beadel. Another
family, the Joneses, maintained similar relationships with Beadel. Alonzo
and Mamie Jones farmed on rented land while their son, Richard, worked
intermittently as a day laborer.[35]

World War II brought sweeping changes to Tall Timbers. For the first
half of the 1940s, Beadel rented land to seven to eight tenants—a significant
number, given that the overall number of black tenants in Leon County fell
by 56 percent during the decade.[36] In an attempt to keep farmers on his land,
Beadel decreased his rental prices in 1941, with rates ranging from $25.00
for a half of a farm to $80.00 for a two-mule farm. Still, farmers contin-
ued to leave.[37] New job and educational opportunities in nearby cities such
Tallahassee and Thomasville drew some, and military service took others
out of the Red Hills. By 1945, the number of contracts had fallen and at the
end of the decade Beadel recorded only two in his ledger. He subsequently
razed some of the small tenant houses and began employing wage laborers
to maintain patch-style agriculture.[38] With the decline of tenancy, day labor
became the norm on Leon County shooting plantations.

MATERIAL CULTURE

Artifacts associated with African American tenants at Tall Timbers offer
insight into the lives of people who left few historical records of their own
and appear most frequently as names in ledgers and public documents.
Surviving artifacts include the remains of houses and outbuildings formerly
used by tenants and materials such as housewares, tools, bottles and jars,
and shoes. Taken together, artifacts suggest that Tall Timbers tenant families
enjoyed modestly better conditions than their counterparts on agricultural
plantations.

Between 2000 and 2005, Morrell and Associates, a private consulting
firm, conducted archaeological investigations at Tall Timbers Plantation with
the assistance of several volunteers and funding from Tall Timbers and the
Archibald Fund. Limited investigations took place between 2000 and 2004.
A grant from the Florida Department of State funded intensive excavations

that took place between August 2004 and April 2005. Morrell and Associates conducted a general surface survey and excavated a sweet potato cellar, a syrup processing area, and a possible refuse pit. These excavations recovered a total of 4,829 artifacts (Table 7.2).

Morrell and Associates' investigations began with examination of extant structures. Beadel razed a large number of buildings in the late 1940s as tenants left Tall Timbers. In the late 1990s, two tenant houses and two corncribs survived.[39] One of these houses and its corresponding corncrib formed the core of a complex that became known as the Jones Tenant House (archaeological site 8Le3536) (Figure 7.1). Researchers focused on this site, which takes its name from its last occupants, Alonzo and Mamie Jones, because the house stood in reasonably good physical condition. Tall Timbers has since rehabilitated it and turned it into an interpretative center. Visitors learn about the lives of African American tenants from five interpretive panels and audio recordings of oral histories with former tenants.[40]

The Jones Family originally occupied a small single-pen structure with an exterior chimney (Figure 7.2). By 1919 it had become a two-room dwelling through construction of an addition. The inhabitant at the time, Josh Forest, lived in the house until his death.[41] Thereafter, Bill and Maggie Jones moved in with their nine children.[42] According to oral histories, no outbuildings stood near the house. The Joneses kept all of their livestock on another farm.[43]

Bill Jones died in 1929. His wife continued farming at Tall Timbers for another year before moving off the plantation. Surviving records offer no clues about the use of the house for the remainder of the decade. By 1942, another family named Jones, unrelated, occupied the structure. From 1942 to 1948 Alonzo and Mamie Lawyer Jones and their ten children made it their home.[44]

Table 7.2   Artifact Concentration Totals

Location	Number of Artifacts
Feature 1—Refuse Pit	3,880
Feature 2—Syrup Processing Area	12
Feature 5	2
Feature 8—Sweet Potato House	369
Feature 9	1
NE Corncrib Stall	139
SW Corncrib Stall	142
Under Crib Floor	172
General Surface Collection	36
Surface Collection—Below the House	49
Above the Back Door in House	4
Smokehouse Test	22
68N47E—Test pit	1

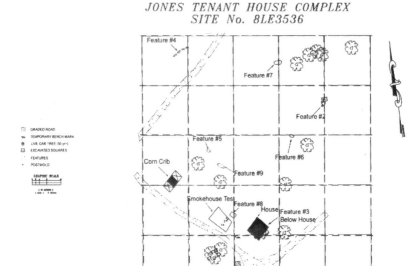

Figure 7.1   Jones Tenant House Site Map. *Source*: Courtesy of author.

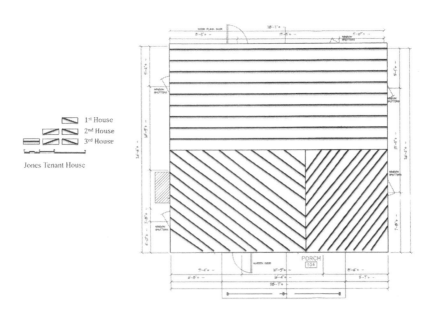

Figure 7.2   Jones Tenant House Plan. *Source*: Courtesy of author.

The Joneses added two rooms to the rear of the house, an interior kitchen and a bedroom.[45] This addition allowed the family to use the new bedroom for their male children, another interior room for female children, and the remaining room—that of the original single-pen structure—as a combination living room and parent's bedroom.[46] The Joneses papered interior walls with newspaper for decoration and as a means of blocking drafts in the manner characteristic of sharecroppers and tenants.[47] Remnants of the Joneses' papering remain visible today.

The porch that spans the façade of the house played an important role in the Joneses' lives by providing exterior living space and opportunities for socializing with neighbors.[48] Most former tenants at Tall Timbers speak of time spent socializing on the porch, especially in the summer months when oppressive heat made interior rooms virtually unbearable.[49] In some cases, tenant families also held weddings on porches.[50] The porch became another room of the house and in many cases the center of the tenant farmer's social life.

The overall form and features of the Jones Tenant House are typical of rural housing of the era. Few tenant houses had modern conveniences such as running water and indoor bathrooms. Interior and exterior finishes lacked the refinement and upkeep typical of in-town dwellings.[51] Tenant houses at Tall Timbers did feature well-built chimneys, however. Most tenant houses of the era had stick and mud chimneys, a simple and inexpensive variety.[52] The Jones House has two brick chimneys: one for the kitchen, the other for an interior fireplace. Beadel viewed brick chimneys as a worthwhile investment since they lessened the potential for fire to damage his property.[53]

Tenant farms at Tall Timbers typically had small groups of outbuildings set near the farmhouse. The number, use, and quality of the outbuildings varied depending on the size of the farm. Oral histories and archaeological surveys have identified buildings that are no longer extant. During its period of use, the outbuildings at the Jones farm included a smokehouse, a sweet potato cellar, and a syrup processing area. Most tenant farms also had a corncrib, which farmers used for storing implements and feed crops. The Jones Tenant House had a corncrib at one time but recollections vary about its dates of use.[54]

Artifacts recovered during fieldwork carried out in 2004 and 2005 came from locations across the Jones Tenant Farm. Classification began with consideration of number of artifacts recovered. The largest concentrations of artifacts came from the refuse pit (Feature 1) and the sweet potato cellar (Feature 8). That these locations yielded large numbers of artifacts is not surprising. Refuse pits and their cousins, privies, are typically artifact rich. It should be noted that materials found in such features might be misleading because they are items that residents no longer wanted. Still, refuse pits and privies contain detritus of everyday life. The artifacts they yield are therefore valuable.[55]

As a site of agricultural labor and storage, the corncrib produced a large number of artifacts. Other portions of the site produced artifact counts below one hundred (Table 7.2). The small number of artifacts found around and below the house likely reflects the practice of "sweeping the yard." Former residents of Tall Timbers recall sweeping the yard around the house with corn shuck broom, a practice common among tenants and sharecroppers in the rural South.[56] Sweeping moved artifacts away from the site of deposition. Although provenience is a central tenet of archaeological research, studies have shown that it does not destroyed sites but merely changes them.[57]

The artifacts recovered from the site reveal a great deal about the tenant farming families. Material groups included glass, metal, ceramics, leather and rubber, plastic, chert, faunal, and other (Table 7.3).

Glass proved most common and came from multiple sources. Artifact colors included blue, green, white, pink, and purple. Many were containers that historically held medicine or foodstuffs. Of the identifiable containers recovered, the majority were fruit jars, with brands such as Mason, Atlas, and Kerr represented.[58] Lids for Atlas-brand jars included wire bail or "wire side" glass lids. Examples of the Kerr brand possessed "Wide Mouth" and self-sealing lids. The most frequent lid recovered was the zinc screw lid with white glass "Boyd's Pat. Liner" (Figure 7.3).[59] Various extract bottles were also identified. Soda bottles, bearing names such as Coca-Cola, Pepsi, Orange Crush, and Nu-Grape were also recovered. The Orange Crush bottle had an embossed patent date of 1920.

Excavations also recovered a small amount of pink Depression glass. Several pieces, including a plate, goblet, drinking glass, and platter, are of a pattern known as Sharon Cabbage Rose (Figure 7.4). Manufactured by the Federal Glass Company from 1935 to 1939, Sharon Cabbage Rose is considered one of the most durable forms of Depression glass. It proved exceedingly popular and was sold by retailers across the country.[60] Circumstantial evidence suggests that the Joneses purchased the Cabbage Rose glass for themselves rather than acquiring it from the Beadel family. Members of the

Table 7.3   Artifact Materials

Material	Number of Artifacts
Glass	3,579
Metal	850
Ceramics	325
Leather and Rubber Composite	51
Plastic	9
Chert	8
Other	5
Faunal	2

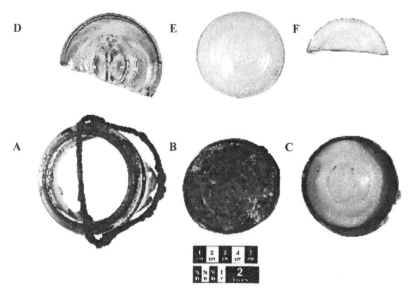

Figure 7.3   Fruit Jar Lids and Liners. A and D, Wire bail style top and glass lid; B, Zinc cap; C, Zinc cap with unidentified glass liner; E, "Genuine Boyd's" cap liner; F, Boyd's variant "Genuine Porcelain" cap liner. *Source*: Courtesy of author.

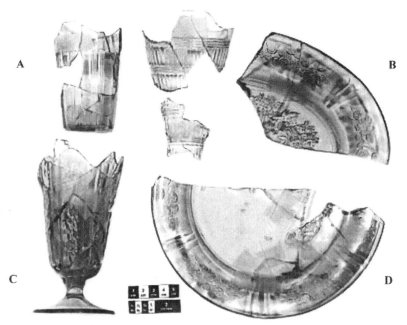

Figure 7.4   Depression glass. A, Tumblers; B, Sharon Cabbage Rose platter; C, Sharon Cabbage Rose goblet; D, Sharon Cabbage Rose plate. *Source*: Courtesy of author.

Jones family recall their mother owning it from the time they were small children. Moreover, the date of manufacture predates the Jones family's move to Tall Timbers.

Moroline bottles are among the most prevalent types of medicinal glass found at the site. Moroline was a fat-based ointment marketed as a hair tonic. African Americans also used it as a first aid ointment and skin moisturizer.[61] Excavations also recovered medicinal bottles embossed with names such as "Sloane's Liniment," "Rawleigh's," and "Fletcher Castoria." All figured among the most common varieties of patent medicines used by rural people during the late 1800s and early 1900s (Figure 7.5).

Agricultural tools constitute the majority of metal artifacts recovered. These tools vary from shallow plow pieces to remnants of mule harnesses to barbed wire fragments. A 1940s-era Florida truck license plate and an electric automobile horn part were also found at Feature 1. No surviving members of the Jones family recall owning an automobile of any kind, but other tenants did.[62]

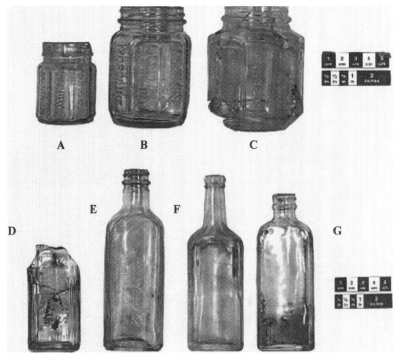

Figure 7.5 Glass containers. A and B, Moroline; C, Lander; D, Sloan's Liniment; E, Rawliegh; F and G, Fletcher Castoria. *Source:* Courtesy of author.

Research also uncovered a large number of shoes and shoe parts, some adult-sized, others for children. Shoe pieces included leather uppers and rubber soles and were primarily found under the house. The protection afforded by the house and the durability of the materials likely accounts for the survival of these items.

Small numbers of other artifacts were found across the site, including eight artifacts made of plastic. Six were buttons found in Feature 8, the corncrib, and during surface collection. A plastic ladies hair comb made of celluloid, a semi-synthetic plastic used since the late 1860s, was found below the house. A few ceramic pieces were also recovered from the site, most of them whiteware, an inexpensive variety common during the period that the Joneses lived at Tall Timbers.[63]

MAKING SENSE OF MATERIAL CULTURE
AT TALL TIMBERS

Interpreting the lives of the people who once lived and farmed Tall Timbers is no easy task. The limitations of the historical record preclude development of a detailed portrait of life and labor. Still, records associated with Tall Timbers tenants offer considerable insight into their existence, and oral histories and the artifacts recovered from the Jones Tenant House allow for more fulsome analysis. Taken together, these sources provide the basis for important conclusions about tenancy on Henry Beadel's shooting plantation.

Archaeologist Charles Orser's study of the Millwood Plantation in Abbeville County, South Carolina, employs a typology based on function rather than material or concentration.[64] This typology assumes that tenant families used materials in ways that would be recognizable to contemporary eyes. Orser uses the example of a small, round, flat object with two holes, also called a button, to illustrate this notion. Just as tenant farmers used buttons as clothing fasteners, so do present-day people. This typology is not perfect—a button could be used as a game marker, for example—but, in general, the parallels are illuminating.[65]

The highest count of artifacts recovered from the Jones Tenant House site, based on the functional typology, are in the categories of foodways and unknown (Table 7.4). Two considerations merit attention. First, many of the artifacts in this category are fragments, which effectively increases the number of items represented. Second, many of the glass artifacts recovered are clear, unmarked fragments, which makes identification of the original containers difficult. Clear glass bottles commonly held medicine, foodstuffs, and a host of other items during the early twentieth century. Yet even when allowances are made for fragmentation and the difficulty of container

Table 7.4   Jones Tenant House Typology Totals

Typology	Number of Artifacts
Foodways	2,191
Unknown	1,635
Labor	578
Household	194
Personal	170
Clothing	61

identification, artifacts used in procuring, preparing, serving, and storing food are most prevalent.

The dominance of the foodways category prompted closer examination of constituent artifacts. Although some are metal—iron stove fragments and cutlery, for example—the single largest group is made up of fruit jar fragments. This is consistent with a large volume of foodstuffs produced on-site. Tenant families used fruit jars mainly for long-term food storage. Use as drinking vessels and for storing non-edible items also occurred but was less common.

The prevalence of artifacts associated with food storage at Jones Tenant House suggests that families on sporting plantations enjoyed somewhat better diets than many of their counterparts. Studies of tenant farmers emphasize dietary monotony and general poor living conditions.[66] Families at Tall Timbers sought self-sufficiency to the best of their ability. They lived off crops and livestock they raised as much as possible. Production of foodstuffs for domestic consumption figures prominently in the recollections of many former residents. Women, for example, recall helping their mothers to preserve fruits and vegetables grown on their farms.[67]

Brand name products such as Sloan's Liniment, Coca-Cola, and the Sharon Cabbage Rose Depression glass demonstrate the reach of consumer goods into the lives of tenants at Tall Timbers. These afford striking evidence of growth of the mass market and its role in the lives of people near the bottom of southern society.[68] Oral histories provide recollections of tenant farm families traveling to nearby cities such as Tallahassee and Thomasville to purchase food, clothing, and other goods. The material record corroborates such recollections. The presence of common consumer goods at Tall Timbers indicates that tenants had sufficient cash to purchase these items on at least some occasions.[69]

Tenants also benefited from events that Beadel sponsored. Each March, Beadel held a rabbit hunt for owners of neighboring plantations, his manager, and his tenants. Former residents recalled these hunts fondly because of the game they yielded and because Beadel gave tobacco to participants.[70] Beadel also sponsored an annual May 20 celebration in commemoration of Emancipation Day in the Red Hills, a notable gesture during an era when white southerners willfully resisted celebrations of African American history. Beadel allowed workers to take the day off and threw a party for his workers and tenants and African Americans from neighboring areas.[71] Owners of other sporting plantations held similar events from time to time. The owners of Pebble Hill Plantation near Thomasville, for example, celebrated Easter and other holidays in comparable fashion.[72] Landowner-sponsored celebrations are virtually unheard of on agricultural plantations.[73] Tenants generally showed limited respect toward landowners on agricultural plantations, while residents of sporting plantations such as Tall Timbers and Pebble Hill recall a general sense of pride in their place of employment.[74]

CONCLUSION

The case of tenancy at Tall Timbers begs questions about other sporting plantations. Did other tenants have similar experiences? What conditions existed on other sporting estates? Although Tall Timbers offers an instructive example, its representativeness is unclear. Circumstances suggest that other landowners had reason to treat their tenants in a similar fashion. Owners of sporting estates, after all, possessed substantial wealth. Whereas owners of agricultural plantations sought profits and had strong incentives to limit costs, owners of sporting estates had reasons to approach tenancy differently. Many likely viewed the rents paid by tenants as useful for offsetting operating costs, and having a pool of potential laborers close at hand also proved valuable. Moreover, the ecological benefits of tenancy became clear over time. Even before Stoddard's research, some landowners recognized that quail coveys tended to favor areas near tenant farms and drew inferences about the relationship between environmental conditions and the birds' behavior. As Stoddard's research explained why quail populations had plummeted in the early 1920s, the value of patch-style farming became clear. Owners of sporting plantations thus had reason to at least limit the coercion that characterized landlord-tenant relations elsewhere, and some may have become more favorably deposed to tenants' presence as knowledge of land-management practices grew.

Tenants on shooting plantations lived difficult lives. Although oral histories tend to recall close-knit communities and a strong sense of pride, it is impossible to overlook the meager circumstances tenants endured, the long hours they worked, the minimal returns they earned, and their place in an economy and social order that marginalized them and offered little hope of advancement. Although some tenants appear to have found tolerable circumstances at Tall Timbers, others did not. The rate of turnover provides a powerful counterpart to the longer periods of residence that some tenants maintained. Contemporary commentators equated turnover with shiftlessness and indolence, but historians have demonstrated the power of movement as a form of resistance to oppression. That roughly half of all tenants who rented land from Beadel stayed two years or less demonstrates the limits of his accommodating stance. Although some tenants preferred Tall Timbers to other plantations, just as many did not.[75]

Still, tenants who stayed at Tall Timbers for several years or more apparently found conditions that allowed for reasonable sustenance and acceptable social relations. The celebrations that Beadel hosted and his modification of contracts and rents are evidence of efforts to treat tenants with dignity and respect, two qualities not commonly associated with black-white relations in the rural South. Beadel's efforts to accommodate tenants' circumstances and market conditions suggests a degree of humanity that other landowners lacked. In an era when many whites viewed African Americans as less than human and thought nothing of cheating them out of payments or using physical violence to achieve their aims, Beadel's actions are significant. They show a disposition to tenants that many of his counterparts lacked.

African Americans left Tall Timbers for the same reasons that tenants and sharecroppers left other plantations—for better opportunities. Beadel's accommodations did not dramatically change the lives of people who earned and had little. His shift to wage labor and mechanized farming is telling evidence of his inability to sustain tenancy at Tall Timbers. As tenants left, Beadel found other measures necessary to maintain the patch-style farming conducive to quail hunting, which resulted in greater outlays of money and effort.[76] Jones family members recall their departure as inspired by the promise of better jobs and educational opportunities in nearby cities.[77] In seeking both, they followed a common path.

Tall Timbers' interpretation of the Jones Tenant House has brought needed attention to an underappreciated dimension of Red Hills history. The exhibit at the house opened in May 2008 and has since become popular with visitors. In February of each year, Tall Timbers hosts a reunion of former tenant families and their descendants. The sight of families gathered and sharing stories of life at Tall Timbers is moving and inspiring, and provides an extraordinary reminder of the historical presence of tenant farming on the southern landscape. From the 1870s to the 1960s, tenant farms were common features across much of the South. Few remain today. Stories passed down through the generations, artifacts, and surviving edifices offer reminders of tenants' experiences and the nation's failure to address debilitating social and economic problems. For all the memories that former residents and descendants share during the February gatherings, it is impossible to overlook the material realities of tenants' lives and their place in a society that emancipated 4.5 million men, women, and children but then turned away from their plight. The failure to aid freedpeople in securing social, economic, and political resources in the aftermath of the Civil War stands among the greatest tragedies in American history.

Ultimately, the example of Tall Timbers shows the need for further research on tenant farming on sporting plantations. Scholars such as Robert Tracy McKenzie and Nancy Virts have argued forcefully for the need to account for temporal and spatial variations in tenancy and sharecropping and the limitations of viewing the as static, homogeneous arrangements. The history of tenancy at Tall Timbers underscores this point. Although historians generally equate tenancy with plantations operated as sites of commercial agriculture, other varieties existed. Excluding sporting plantations from the picture obscures vital histories and results in an impoverished understanding of tenancy, how it changed over time, and why important changes occurred.[78]

NOTES

1. Gilbert Fite, *Cotton Fields No More: Southern Agriculture, 1865–1980* (Lexington: University Press of Kentucky, 1984), chap. 1; Charles S. Aiken, *The Cotton Plantation South Since the Civil War* (Baltimore: Johns Hopkins University Press, 1998), chaps. 1 and 2; Roger L. Ransom and Richard Sutch, *One Kind of Freedom: The Economic Consequences of Emancipation* (New York: Cambridge University Press, 1977), chaps. 4 and 5; Lee J. Alston and Kyle D. Kauffman, "Agricultural Chutes and Ladders: New Estimates of Sharecroppers and 'True Tenants' in the South, 1900–1930," *Journal of Economic History* 57, no. 2 (June 1997): pp. 464–475.

2. Fite, *Cotton Fields No More*, pp. 38, 161–162; Frank Tannenbaum, *Darker Phases of the South* (New York: G. P. Putnam's Sons, 1924), p. 133; Natalie J. Ring, *The Problem South: Region, Empire, and the New Liberal State, 1880–1930* (Athens: University of Georgia Press, 2012), chap. 3.

3. Tall Timbers Plantation (Leon County, Fla.), National Register of Historic Places nomination, Florida Department of State, Tallahassee, Fla.; Herbert L Stoddard, *The Bobwhite Quail: Its Habits, Preservation, and Increase* (New York: Charles Scribner and Sons, 1931); Albert G. Way, *Conserving Southern Longleaf: Herbert Stoddard and the Rise of Ecological Land Management* (Athens: University of Georgia Press, 2011); "Welcome to Tall Timbers Research Station and Land Conservancy," http://talltimbers.org/welcome-to-tall-timbers/.

4. Leon County Land Records, 1919, Deed Book 1, pp. 225–228, Leon County Courthouse, Tallahassee, Fla. (hereafter LCC).

5. Tall Timbers Plantation, National Register of Historic Places nomination; "Welcome to Tall Timbers Research Station and Land Conservancy," http://talltimbers.org/welcome-to-tall-timbers/.

6. Certificate of the Land Office no. 538, 1826, LCC; Leon County, Fla., *Sixth Census of the United States, 1840* (Microfilm M704), U.S. Bureau of the Census,

Record Group 29, National Archives and Records Administration, Washington, D.C. (hereafter NARA); Clifton Paisley, *From Cotton to Quail: An Agricultural Chronicle of Leon County, Florida, 1860–1967* (Gainesville: University of Florida Press, 1968), p. 127.

7. Clay Ouzts, "Landlords and Tenants: Sharecropping and the Cotton Culture in Leon County Florida, 1865–1885," *Florida Historical Quarterly* 75, no. 1 (summer 1996), p. 2.

8. Ouzts, "Landlords and Tenants," pp. 1–23; *Semi-Weekly Floridian* (Tallahassee, Fla.), Jan. 4, 1867, quoted in Ouzts, "Landlords and Tenants," p. 3; Susan Hamburger, "On the Land for Life: Black Tenant Farmers on Tall Timbers Plantation," *Florida Historical Quarterly* 66, no. 2 (Oct. 1987): pp. 152–159; Paisley, *From Cotton to Quail*, pp. 22–29; Alston and Kauffman, "Agricultural Chutes and Ladders," pp. 465–466.

9. Deed of sale, Griffin Holland to Alexander Moseley, 1871, Deed Book P, pp. 612–613, LCC.

10. Deed of sale, Alexander Moseley to Eugene H. Smith, 1880, Deed Book W, pp. 143–144, LCC.

11. Leon County, Fla., *Ninth Census of the United States, 1870* (Microfilm M593), Bureau of the Census, NARA.

12. Deed of sale, Elizabeth Smith to Edward Beadel, 1895, Deed Book FF, pp. 142–143, LCC; Tall Timbers Plantation, National Register of Historic Places nomination.

13. Tall Timbers Plantation, National Register of Historic Places nomination.

14. Ibid.

15. Henry Beadel Business Ledger, 1920–1952 (hereafter Beadel Ledger); Annie Bell Sloan, interview by Juanita Whiddon, May 11, 2005, tape recording; both at Tall Timbers Research Station Archives, Tallahassee, Fla. (hereafter TTRS).

16. Tall Timbers Plantation, National Register of Historic Places nomination.

17. Beadel did not record the size of the farms rented, only the rental costs. Acreages can be surmised from oral interviews and notes found in his personal ledger. See, for example, Beadel Ledger; Annie Bell Sloan interview.

18. Enumeration District 99, Meridian, Leon, Fla., *Fourteenth Census of the United States, 1920* (Microfilm T625), U.S. Bureau of the Census, Record Group 29, NARA.

19. Beadel Ledger.

20. Florida Crop and Livestock Reporting Service, *Florida Agricultural Statistics, Field Crop Data 1919–1967* (Orlando: n.p., 1967), pp. 3, 7.

21. Alex and Hattie Sloan, interview by Juanita Whiddon, July 27, 2005, tape recording, TTRS.

22. Paisley, *From Cotton to Quail*, p. 38.

23. Harry F. Brubaker, "Land Classification, Ownership, and Use in Leon County, Florida" (Ph.D. diss., University of Michigan, 1956), p. 102.

24. Beadel Ledger.

25. Enumeration District 99, Meridian, Leon County, Fla., *Fourteenth Census of the United States, 1920* (Microfilm T625), U.S. Bureau of the Census, Record Group 29, NARA.

26. Enumeration District 0001, Meridian, Leon County, Fla., *Fifteenth Census of the United States, 1930* (Microfilm T626), U.S. Bureau of the Census, NARA.

27. Precinct Number 8, Leon County, Fla., Eleventh Census of the State of Florida, 1945 (Microfilm S1371), Record Group 001021, State Library and Archives of Florida, Tallahassee, Fla.

28. Charles S. Johnson, *Shadow of the Plantation* (Chicago: University of Chicago Press, 1934), p. 32.

29. Beadel Ledger.

30. Johnson, *Shadow of the Plantation*, p. 117.

31. E. A. Goldenweiser and Leon E. Truesdell, *Farm Tenancy in the United States* (Washington, D.C.: Government Printing Office, 1924), pp. 135–136. The South Atlantic region included Delaware, Maryland, the District of Columbia, Virginia, West Virginia, South Carolina, North Carolina, Georgia, and Florida.

32. Beadel Ledger.

33. Lonnie A. Marshall, "The Present Function of Vocational Agriculture, Schools and Other Agencies in the Rural Improvement of Negroes in Leon County, Florida" (M.A. thesis, Iowa State College, 1930), pp. 105, 108, 120, quoted in Paisley, *From Cotton to Quail*, p. 104.

34. Beadel Ledger.

35. Ibid.

36. Brubaker, *Land Classification*, p. 102.

37. Beadel Ledger.

38. Tall Timbers Plantation, National Register of Historic Places nomination.

39. Ibid.

40. On the preservation and interpretation of the Jones House, see "The Jones Family Tenant Farm," Tall Timbers Plantation, http://talltimbers.org/the-jones-family-tenant-farm/.

41. Annie Bell Sloan interview.

42. Enumeration District 99, Meridian, Leon County, Fla., *Fourteenth Census of the United States, 1920* (Microfilm T625), U.S. Bureau of the Census, Record Group 29, NARA.

43. Annie Bell Sloan interview.

44. Beadel Ledger.

45. Minnie Jones Leonard, interview by Juanita Whiddon, Sept. 8, 2004, tape recording, TTRS.

46. Rosalie Jones Brim, interview by Juanita Whiddon, June 26, 2004, tape recording, TTRS.

47. Johnson, *Shadow of the Plantation*, p. 92.

48. Aiken, *The Cotton Plantation South*, pp. 135–137.

49. Leonard interview.

50. Annie Bell Sloan interview.

51. Johnson, *Shadow of the Plantation*, p. 92.

52. Charles E. Orser, Jr., *The Material Basis of the Postbellum Tenant Plantation* (Athens: University of Georgia Press, 1988), p. 94.

53. Juanita Whiddon, conversation with author, Sept. 2004.

54. Annie Bell Sloan interview.

55. Charles E. Orser, Jr., and Brian M. Fagan, *Historical Archaeology* (New York: Harper Collins College Publishers, 1995), p. 61.

56. Annie Bell Sloan interview.

57. Orser, *Material Basis*, p. 100.

58. Dave Hinson, "A Primer on Fruit Jars," accessed on June 28, 2014, http://home.qnet.com/~glassman/newsletter/primer.pdf.

59. L. Ross Morrell and Robin T. Bauer, "Jones Tenant Farm 8Le3536: Archaeological and Historical Investigations" (Bureau of Historic Preservation, Florida Division of Historical Resources, Florida Department of State, 2005), p. 23.

60. Gene Florence, *The Collector's Encyclopedia of Depression Glass*, 7th ed. (Paducah, Ky.: Collector Books, 1986), p. 184.

61. Laurie A. Wilkie, *Creating Freedom: Material Culture and African American Identity at Oakley Plantation, Louisiana 1840–1950* (Baton Rouge: Louisiana State University Press, 2000), pp. 174–175.

62. Alex and Hattie Sloan interview.

63. "Evolution of English Household Tableware: Period VI: 1820s-1900," Nautical Archaeology at Texas A&M, accessed July 15, 2014, http://nautarch.tamu.edu/class/313/ceramics/period-6.htm.

64. David H. Chance, review of *Method and Theory in Historical Archeology*, by Stanley South, *Historical Archaeology* 11 (1977), pp. 126–128.

65. Orser, *Material Basis*, p. 234.

66. For example, Arthur Raper noted that wage workers in Greene and Macon County, Georgia, spent half of their earned income on food, while sharecroppers and renters spent on average ten dollars more than half. Such expenditures suggest that food was obtained through purchase, not through stored goods used over time. Raper, *Preface to Peasantry: A Tale of Two Black Belt Counties* (Chapel Hill: University of North Carolina Press, 1936). See also Fite, *Cotton Fields No More*, p. 34.

67. Leonard interview.

68. Fite, *Cotton Fields No More*, p. 34.

69. Alex and Hattie Sloan interview.

70. Emmit and Celia Gay, interview by Juanita Whiddon, Feb. 12, 2004, tape recording, TTRS.

71. Brim interview.

72. Princetta Hadley Green, interview by Jack Hadley, in Titus Brown and James "Jack" Hadley, *African American Life on the Southern Hunting Plantation* (Charleston, S.C.: Arcadia Press, 2000), pp. 103–105.

73. Jack Temple Kirby, *Rural Worlds Lost: The American South 1920–1960* (Baton Rouge: Louisiana State University Press, 1987), p. 142.

74. Sam Green, interviewed by Jack Hadley, in *African American Life*, p. 46.

75. On movement as a form of resistance, see Jacqueline Jones, *The Dispossessed: America's Underclasses from the Civil War to the Present* (New York: Basic Books, 1992), chap. 4.

76. Tall Timbers Plantation, National Register of Historic Places nomination.

77. Brim interview.

78. Robert Tracy McKenzie, "Freedman and the Soil in the Upper South: The Reorganization of Tennessee Agriculture, 1865–1880," *Journal of Southern History* 59, no. 1 (Feb. 1993): pp. 63–84; Robert Tracy McKenzie, *One South or Many?: Plantation Belt and Upcountry in Civil War-Era Tennessee* (New York: Cambridge University Press, 1994); Nancy Virts, "Change in the Plantation System: American South, 1910–1945," *Explorations in Economic History* 43, no. 1 (Jan. 2006): pp. 153–176.

Index

About the Contributors

Robin Bauer Kilgo is a museum consultant in the Florida Keys. Her recent work includes projects for the Florida Association of Museums and the Association of Registrars and Collections Specialists. She previously served as Registrar and Collections Officer for the Seminole Tribe of Florida's Ah-Tah-Thi-Ki Museum. Kilgo received her M.A. in Historical Administration and Public History from the Florida State University in 2005. Her thesis, "At Home Among the Red Hills: The African American Tenant Farm Community on Tall Timbers Plantation," examines the lives of African American tenants on one of the largest shooting plantations in the Red Hills.

Jennifer Betsworth is a historic preservation specialist at the New York State Office of Parks, Recreation, and Historic Preservation. She earned an M.A. in Public History from the University of South Carolina in 2011 and is the author of several National Register of Historic Places nominations. Her thesis, "'Then Came the Peaceful Invasion of the Northerners': The Impact of Outsiders on Plantation Architecture in Georgetown County, South Carolina," examines changes in architecture and landscape that accompanied the development of northern-owned estates in Georgetown County, South Carolina, during the early twentieth century.

Julia Brock is assistant professor of history and co-director of the Center for Public History at the University of West Georgia. She received a Ph.D. in Public History from the University of California, Santa Barbara. Her dissertation, "Land, Labor, and Leisure: Northern Tourism in the Red Hills Region, 1890–1950," examines the development of a northern hunting colony in the Red Hills region, a forty-mile stretch of land in southwest Georgia and northern Florida, and its impact on land use, African American labor, and game laws in the state of Georgia.

Matthew A. Lockhart is the editor of the *South Carolina Historical Magazine* and a Ph.D. candidate in history at the University of South Carolina. His dissertation, "From Rice Fields to Duck Marshes: Sport Hunting and Environmental Change on the South Carolina Coast, 1900–1950," examines the rise of recreational hunting and its environmental consequences in the South Carolina lowcountry during the first half of the twentieth century. He has published articles in the *Public Historian*, *Southeastern Geographer*, and the *South Carolina Historical Magazine*.

Hayden Ross Smith received his Ph.D. in history from the University of Georgia in 2012. His dissertation, "Rich Swamps and Rice Grounds: The Specialization of Inland Rice Culture in the South Carolina Lowcountry, 1670–1861," examines the interplay of technology, culture, and the environment in the development of commercial rice production in coastal South Carolina. His article "Reserving Water: Environmental and Technological Relationships with South Carolina Rice Plantations" appears in *Rice: Global Networks and New Histories*, ed. Francesca Bray, Peter Coclanis, Edda Fields-Black, and Dagmar Schafer (Cambridge University Press, 2015). He currently teaches at the College of Charleston.

Drew Swanson is an assistant professor of history at Wright State University. He earned a Ph.D. from the University of Georgia in 2010 and taught at Millsaps College in Jackson, Mississippi, from 2011 to 2013. He is the author of *A Golden Weed: Tobacco and Environment in the Piedmont South* (Yale University Press, 2014) and *Remaking Wormsloe Plantation: The Environmental History of a Lowcountry Landscape* (University of Georgia Press, 2012).

Daniel Vivian is an assistant professor of history at the University of Louisville. A specialist in the history of the American South, he received his Ph.D. from the Johns Hopkins University in 2011. His writings have appeared in *Winterthur Portfolio*, the *South Carolina Historical Magazine*, and *Ohio Valley History*. He is currently revising his dissertation, "The Leisure Plantations of the South Carolina Lowcountry, 1900–1940," for publication.

Lightning Source UK Ltd.
Milton Keynes UK
UKOW02n0411250217

295309UK00010B/161/P